Scotland's leading educational publishers

CfE Higher
GEOGRAPHY
GRADE A BOOSTER

CfE Higher GEOGRAPHY *GRADE A BOOSTER*

Carly Smith

001/22042015

10 9 8 7 6 5 4 3

ISBN 9780007590896

Published by
Leckie & Leckie Ltd
An imprint of HarperCollins*Publishers*
Westerhill Road, Bishopbriggs, Glasgow, G64 2QT
T: 0844 576 8126   F: 0844 576 8131
leckieandleckie@harpercollins.co.uk
www.leckieandleckie.co.uk

Publisher: Katherine Wilkinson
Project manager: Craig Balfour

Special thanks to
Louise Robb (copy edit and proofread)
Keren McGill (proofread and editorial)
Jouve (layout)
Ink Tank (cover)

Printed and bound by CPI Group (UK) Ltd,
Croydon, CR0 4YY

A CIP Catalogue record for this book is available from the British Library.

Acknowledgements

All images © Shutterstock.com

Illustrations © HarperCollins Publishers

This product uses map data licensed from Ordnance Survey © Crown copyright and database rights (2014) Ordnance Survey (100018598). Pages 39, 40, 41, 42, 133, 134, 141 and 142.

Contents

Section 1: Introduction

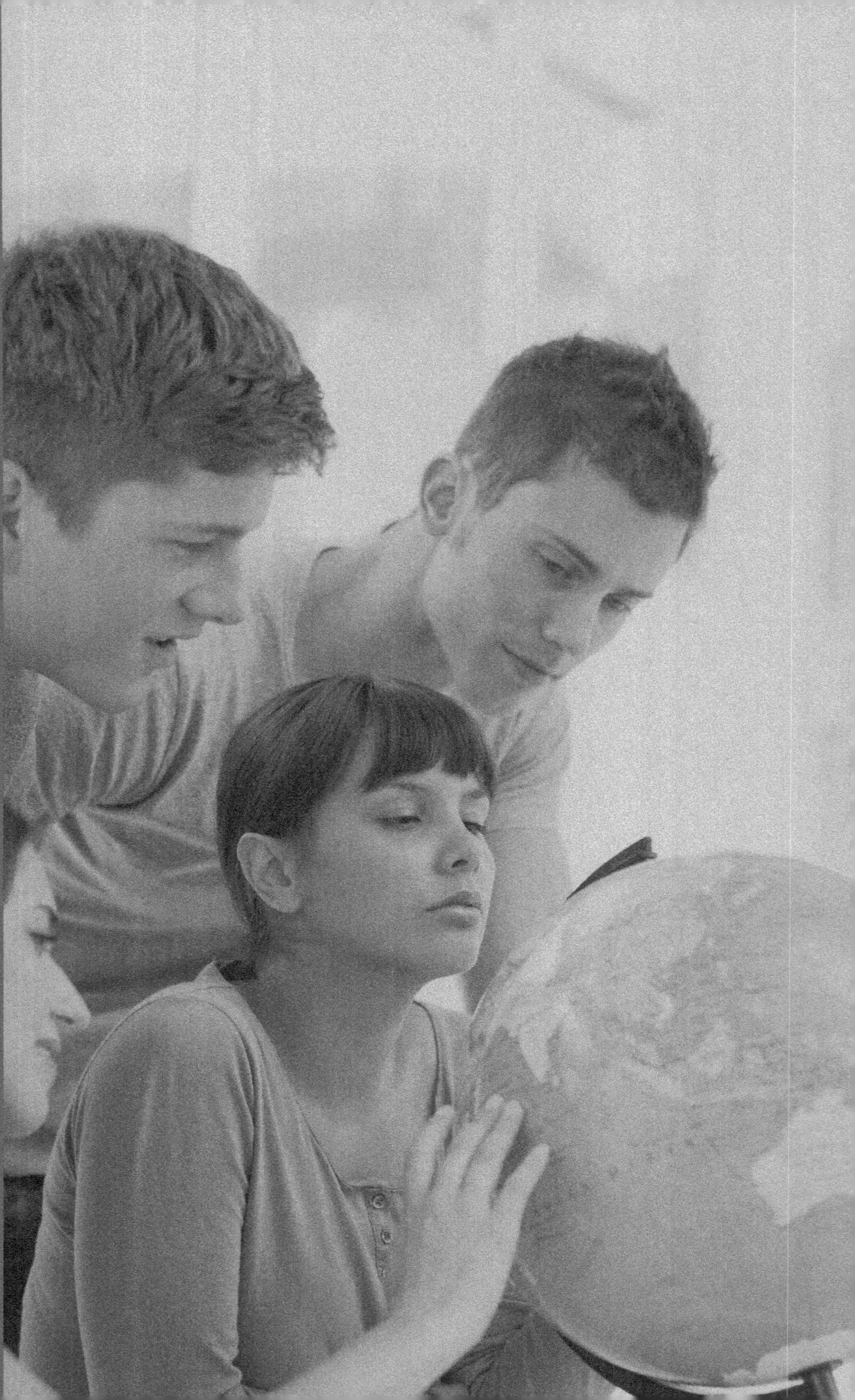

Introduction

Layout of the book

This book covers each of the four sections within the examination paper and gives advice on completing the course Assignment.

Within each section of the book there will be a topic summary, sample questions, worked examples and a glossary. Before reading over the sample answers, try each of the questions for yourself. This should help you identify any gaps in your knowledge as well as any flaws with your technique. It is important to remember this book is not a coursework guide and should be used in conjunction with your course notes.

Maximising marks

The 'Maximising Marks' section within each section of this book is there to help you to identify where marks are awarded and how much detail you need to include. Understanding the standards required will save you time and can help you pick up those vital marks.

External examination – a brief overview

The external examination is made up of one paper which lasts 2 hours 15 minutes. The paper is divided into four sections, totaling 60 marks.

Section 1 – Physical Environments (15 marks)
Topics include **Atmosphere, Biosphere, Lithosphere** and **Hydrosphere.** You will be asked three questions that will cover any of these four topics. There is no choice – you should attempt all three questions in section one.

Section 2 – Human Environments (15 marks)
Topics include **Population, Urban** and **Rural**. There is no choice – you should attempt all three questions in section two.

Section 3 – Global Issues (20 marks)
Section three is made up of five questions from which you will answer two. Topics include **River Basin Management, Development and Health, Global Climate Change, Trade, Aid and Geopolitics** and **Energy**. You should answer the questions that relate to the units you have studied in class. Each question is marked out of 10.

Section 4 – Application of Geographical Skills (10 marks)

Section four contains one question based on an Ordnance Survey map and a number of other sources. You will be given a scenario in which you should use the sources to help you reach a conclusion.

The Assignment

The Assignment is worth 30 marks, which is 33% of your overall award. This should be based on primary sources such as fieldwork or secondary sources such as the internet. The write-up stage of the Assignment will be completed under exam conditions in 1 hour 30 minutes. See section 15 of this book for more guidance on this.

Spice up your revision

Add some colour

Visual aids can be helpful when revising. Drawing colourful mind maps, pictures or tables will help you to memorise facts. You could also use colour to classify items, e.g. long/short term effects, effects on people/environment. Pinning these notes to your wall will help reinforce your knowledge as the big day creeps closer.

Know the key words

Make a list of key words for each section. Again, this could be colour coded. Knowing key words will aid your understanding of each topic and will help you gain marks in the exam.

Identify gaps

When revising any topic, challenge yourself to write down everything you already know then highlight where the gaps lie. Be disciplined with yourself and fill these knowledge gaps where necessary. Closer to the exam, condense your revision notes on to small cards. This is particularly useful for case studies.

Complete past papers

Past papers should play a prominent role in your exam preparation. Not only will they help you improve your exam technique but they will also allow you to identify trends in the questions asked.

Top Tip

Use the traffic light system to help you identify any gaps in learning. Look through your notes and highlight topics with red (not got a clue), amber (needs more revision) or green (fully understand). As you progress with your revision, try to turn the red into amber and then the amber into green.

Practice drawing diagrams

If a question specifically asks for annotated diagrams marks will be lost if you do not include them. However, it is a good idea to include diagrams in any answer where appropriate. They can save you time and help you pick up extra marks in the process. The trick with diagrams is to keep them simple. There is no point including a work of art if it leaves you with no time to answer the question.

In the exam

Use all resources provided

Some questions may be based on resources such as maps, diagrams, graphs or tables. It costs the exam board a lot of money to produce these resources – they do it for a reason! Use all resources to your advantage and extract information where appropriate. This could be in the form of grid-references, quotes or figures.

Top Tip

Be careful! When using information from maps, diagrams or graphs – be precise. Make sure your grid references are correct and you include the correct units when asked for measurements.

Read all questions carefully

Before writing your answer you should identify the key words in the question. It may be useful to underline them or use a highlighter. It is also a good idea to stop half way through your answer and read the question again. This will keep you focused and ensure your answer is relevant.

Avoid lists

A simple list or bullet pointed answer can achieve a maximum of half marks. Always write in sentences and develop every point where you can.

Know your case studies

Case study knowledge is a vital part of the Geography exam. At Higher level it is not enough to simply name the area you have studied. Make sure you are familiar with all of your case studies and be able to provide specific named examples and numerical data where appropriate, e.g. tourist attractions or visitor numbers.

If you are completely stuck

Leave the question blank and come back to it at the end. If, after a second glance you are still stuck – check the keywords in the question and write everything you can remember about them. You never know, you might pick up some extra marks. No answer means no marks, so never leave a question blank.

If you are running out of time

Don't panic. Look at the questions you have left to answer and divide up your remaining time to cover them all. Be very economical – make one point, support it with evidence and then move on to the next point. If you really can't finish in time, briefly list the points you wanted to make – they could pick you up a few marks.

Command words:

What are they and why are they important?

There are a range of questions that you could be asked within the question paper, each defined by its own 'command word'. It is important to familiarise yourself with these to ensure you know what the question is asking you to do. Some examples of command words are given below.

Explain

Questions that ask you to **explain** or give reasons for the cause or impact of something require candidates to do more than describe to gain credit. For example, if asked to *explain the problems of collecting accurate population data in developing countries,* you should refer to each factor by stating **why** it makes data collection difficult. A purely descriptive answer, or one where development is limited, will achieve no more than half marks.

Analyse

Analysis involves identifying relationships between different parts/components. Analysis marks are awarded when you use your own **knowledge** and **understanding** to identify relevant components (e.g. of an argument, theory, idea) before showing at least one of the following:

- similarities and contradictions
- different views
- different interpretations
- possible consequences
- links between different components and related concepts

Where you are asked to **analyse**, you should identify parts of the topic/issue and refer to the interrelationships between various factors. For example, *with reference to an international migration you have studied,* **analyse** *the impact on either the donor country or the receiving country.* Here, you are expected to illustrate what effect migration has had on either the donor country or the recipient country.

Evaluate

Where you are asked to **evaluate**, you should be making a judgement of the success, failure, or impact of something based on given criteria. For example, *referring to different countries,* **evaluate** *the suitability of renewable approaches to generating energy.* Here, you are expected to briefly describe the different renewable approaches before commenting on their ability to generate energy.

Account for

When asked to **account for**, you are expected to **give reasons**. For example, **account for** *the economic, environmental and social benefits of a named water control project.* Here, you are expected to suggest reasons for the benefits created by a water control project you have studied.

Discuss

Discuss style questions are looking for you to consider different views on an issue or argument. These views do not necessarily need to be balanced. However, you should include a range of arguments/impacts within your answer. For example, *with reference to Figure 1.1,* **discuss** *the possible consequences of the 2050 population structure for the future economy of Malawi and the welfare of its citizens.* Here, you should talk about the positive **and** negative impacts of population change with reference to both social **and** economic factors.

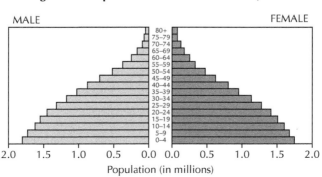

Figure 1.1: Population structure for Malawi, 2050

To what extent

This type of question asks you to consider the impact of a management strategy or strategies you have explored. For example, *with reference to a specific National Park,* **to what extent** *have the negative effects of tourism been tackled by National Park authorities?* Here, you should briefly describe the negative effects of tourism before commenting on the success of the strategies employed.

Top Tip

It is a good idea to highlight/ underline the key word as you read through the question. This will keep you focused and help you answer the question correctly.

Reading a question carefully is a vital part of the exam process. Misreading an instruction on a question can prove very costly to the overall number of marks awarded.

SECTION 2

Atmosphere

Topics include:

- Global heat budget
- Redistribution of energy by atmospheric and oceanic circulation
- Causes and impacts of the Intertropical Convergence Zone

Global heat budget

The global heat budget is the balance between incoming and outgoing solar radiation. You should be able to explain how solar energy varies at different times of the year and for different locations across the globe.

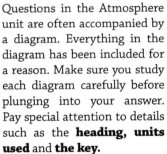

Top Tip

Questions in the Atmosphere unit are often accompanied by a diagram. Everything in the diagram has been included for a reason. Make sure you study each diagram carefully before plunging into your answer. Pay special attention to details such as the **heading, units used** and **the key.**

Q Study Figures 2.1 and 2.2. **Explain** why the Earth's surface absorbs only 50% of the solar energy received at the edge of the atmosphere. You should refer to both conditions in the Earth's **atmosphere** and at the Earth's **surface**.

5 marks

Figure 2.1: Earth/atmosphere energy exchange

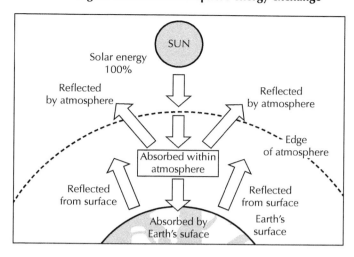

Figure 2.2: Proportion of solar energy absorbed/reflected

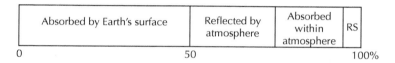

Absorbed by Earth's surface	Reflected by atmosphere	Absorbed within atmosphere	RS

0   50   100%

Read the two answers below. What differences do you notice between them? Think about structure, relevance to the question and attention to detail.

Answer A

The Earth only receives around 50% of the solar insolation that comes from the sun. Some is reflected straight back into space due to the albedo effect. Some is absorbed by clouds, water vapour, dust particles and a number of other gases. **(1 mark)** *So, some energy is reflected by the atmosphere and some is absorbed within the atmosphere. A small amount is also reflected.*

Energy absorbed by the Earth causes the Earth's temperature to rise. Clouds and gases absorb some of this. Most of the energy which actually heats the atmosphere comes from the surface of the Earth.

1 out of 5 marks

Answer B

The Earth's surface only absorbs 50% of solar energy because reflection and scattering reduces the amount of solar energy by 30%. **(1 mark)** *This is known as the albedo effect. This energy is sent back into the atmosphere as long wave radiation.* **(1 mark)**

Approximately 20% is reflected by clouds and 5% is scattered by gas and dust particles whilst 5% is reflected from the Earth's surface. **(1 mark)**

Reflection therefore varies depending on the extent of cloud cover and the covering of the Earth's surface. Snow and ice covered surfaces at the poles reflect more radiation compared to darker forest areas at the tropics which absorb more incoming radiation. **(1 mark)**

Absorption by the atmosphere reduces the solar energy by approximately 20%. This is because 17% is absorbed by dust, gas and water vapour, while 3% is absorbed by clouds. **(1 mark)**

5 out of 5 marks

What answer comes out on top?

Answer A has achieved 1 out of 5. This candidate has relied heavily on the use of the word 'some' when describing the movement of energy instead of taking figures directly from the diagrams provided. This is careless, resulting in an answer that is vague and over-generalised. The 'albedo effect' has been mentioned within the answer. However, the candidate has not explained this point or linked it to the question, therefore failing to pick up the marks. With more precision this could be a good answer but in this instance poor exam technique has let the candidate down.

On the other hand, answer B has achieved 5 out of 5, with the candidate referring to conditions both in the atmosphere and at the Earth's surface. The terms 'reflection' and 'absorption' have both been used well to explain the exchange of energy and the figures given in the diagram have been put to good use. This is a well-structured response with the candidate successfully applying their knowledge to the question.

Redistribution of energy

You should be able to explain, using an appropriate diagram, why tropical latitudes receive more solar energy than polar regions.

Q With the aid of an annotated diagram or diagrams, **explain** why there is a surplus of solar energy in the tropical latitudes and a deficit of solar energy towards the poles. **5 marks**

Answer A

Pupil diagram A: Redistribution of energy

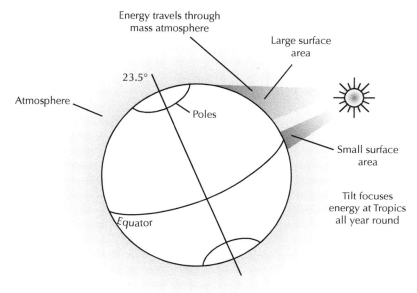

There is a surplus of solar energy in tropical latitudes as this is where the sun's rays are more concentrated. The sun's rays strike the areas around the centre of the earth at right angles therefore the intensity of insolation is greatest. **(1 mark)** There is also less atmosphere to pass through at the tropics. This results in less energy lost due to absorption and reflection by clouds and gas. **(1 mark)**

At higher latitudes the rays strike the surface at a wider angle, which spreads heat energy over a bigger surface area. **(1 mark)** Therefore, the insolation striking the surface at the equator heats up a smaller surface area than the same amount of insolation at higher latitudes, resulting in a surplus of solar energy. **(1 mark)**

The albedo effect also makes a difference. More radiation is absorbed at the tropics due to the vast amount of dark forestry. However, more radiation is reflected at the poles by the amount of ice. **(1 mark)**

5 out of 5 marks

What makes this a good answer?

The candidate has successfully broken up the question and used it to structure their answer. Each point has been developed fully and the albedo effect has been explained well. The diagram has been annotated to show features not already mentioned in the written response, e.g. 'a tilt in the Earth's axis focuses energy at the tropics throughout the year'. This lack of repetition ensures good use of time and would have earned the candidate an additional mark if needed.

Answer B

Pupil diagram B: Redistribution of energy

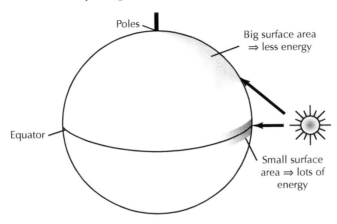

The amount of heat received from the sun varies across the globe. Equatorial areas are always very warm whilst polar areas are cold.

The sun's rays are more intense at the equator and they are concentrated over a smaller area than at the poles. This will make the Earth's surface hotter here. **(1 mark)** You can see this in the diagram above.

At the poles there are alternating six months of light and six months of darkness due to the movement of the earth around the sun. However the tropics receive a lot of insolation throughout the year, allowing that energy to be stored. **(1 mark)**

2 out of 5 marks

What makes this a weak answer?

The first paragraph is just a simple rewording of the question and receives no marks. This is careless and a poor use of time. The second paragraph offers limited description and explanation. However, this candidate uses the diagram to help develop their point resulting in 1 mark being awarded. Lastly, the third paragraph includes two limited explanations earning the candidate an additional mark. Although the candidate has clearly understood the question, their response is basic and opportunities have been lost to further develop points that would have earned this candidate more marks.

Maximising Marks

Always look at the number of marks awarded and write your answer accordingly. In order to achieve one mark you should give one detailed explanation, e.g.

'The insolation striking the surface at the equator heats up a smaller surface area than the same amount of insolation at higher latitudes, resulting in a surplus of solar energy.' **(1 mark)**

OR a limited description/explanation of two factors, e.g.

'The sun's rays are more intense at the equator and they are concentrated over a smaller area than at the poles. This will make the Earth's surface hotter here as the sun's rays have to travel through more atmosphere to reach the poles, losing energy.' **(1 mark)**

Intertropical Convergence Zone (ITCZ)

You should be familiar with the causes and impact of the ITCZ at the thermal equator. To understand the causes and impact you must first be familiar with the trade winds and air masses that occur over Africa.

The north-east trade winds come from the Tropical Continental (cT) air mass whilst the south-west trade winds come from the Tropical Maritime (mT) air mass.

A summary of these two air masses can be seen in the figure below.

Figure 2.3: Characteristics of the Tropical Maritime and the Tropical Continental air mass

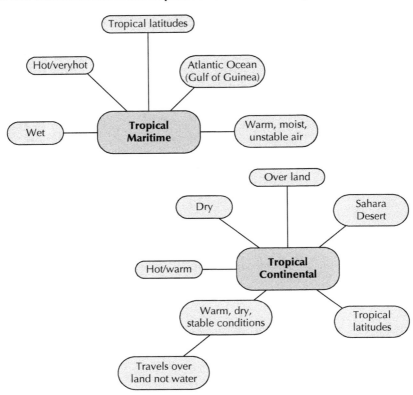

It is not enough to give vague comments, e.g. formed over the sea or formed over the land. For full credit to be given you must name specific sea and land areas, e.g. formed over the Atlantic Ocean or formed over the Sahara Desert.

A typical question on the ITCZ will often rely on you analysing a number of graphs and maps. It is important that you refer to the sources provided and incorporate them into your answer.

Q Study Figures 2.4, 2.5 and 2.6. **Explain** how the ITCZ affects the rainfall patterns in West Africa.
5 marks

Figure 2.4: ITCZ

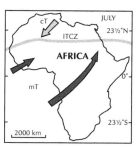

Figure 2.5: West Africa

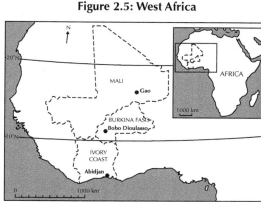

Key:
mT = Maritime Tropical   cT = Continental Tropical
ITCZ = Inter Tropical Convergence Zone

Figure 2.6: Average monthly rainfall/days with precipitation

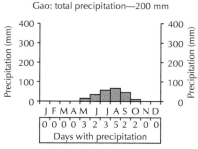

Gao: total precipitation—200 mm

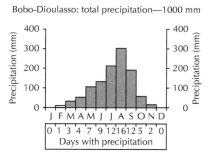

Bobo-Dioulasso: total precipitation—1000 mm

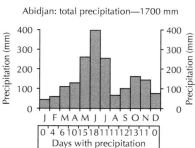

Abidjan: total precipitation—1700 mm

Tackling the question

With a question like this, take each region in turn and use the climate graph provided to **explain** fully the impact of the ITCZ.

1. Look at when the rainfall occurs (Is it all year round? Certain months of the year? Are there distinct wet/dry seasons?) – incorporate this into your answer.
2. Work out total rainfall for the year (approximately) – use this to compare different regions.
3. Quote figures from all graphs provided; credit will be given for this.

Once you have analysed each graph, use the map to help you explain the patterns shown. An example of how to structure your answer is shown below.

For 6 months of the year, Gao will see no rainfall at all. This is because it is under the influence of **hot, dry cT** *air for most of the year.* **(1 mark)** *Therefore, Gao has only 17 rain days and a very low total annual precipitation figure as it lies well to the north of the ITCZ for most of the year.* **(1 mark)**

Compare this now to Abidjan.

Abidjan, which lies on the Gulf of Guinea coast, is influenced by **hot, humid mT** *air for most of the year. This explains the high total annual precipitation figure and large number of rain days.* **(1 mark)** *The twin peaks of precipitation happen because the ITCZ moves northwards in the early part of the year and then south later in the year in line with the overhead sun.* **(1 mark)**

Go on to discuss Bobo-Dioulasso

Bobo-Dioulasso has a **clear wet season/dry season**. *It receives more rain days and heavy precipitation from June–August when the ITCZ is furthest north.* **(1 mark)** *This brings rainfall to the area because the mT air mass is dominant.* **(1 mark)**

Top Tip ✔

When writing your answer, always state the **dominant air mass** and the effect this may have on the area. If air masses are not mentioned, full marks cannot be awarded.

Glossary

Air mass: A large body of air that affects the weather of an area over which it moves.

Albedo: The amount of incoming solar radiation that is reflected by the Earth's surface and the atmosphere.

Coriolis force: The deflection of moving air (winds) produced by the rotation of the Earth on its axis.

Insolation: Heat energy from the sun (**in**coming **sol**ar radi**ation**).

Intertropical Convergence Zone: The place in the atmosphere along which the trade winds meet (converge).

Pressure belts: Patterns of atmospheric circulation systems of either high or low atmospheric pressure.

Section 3: Hydrosphere

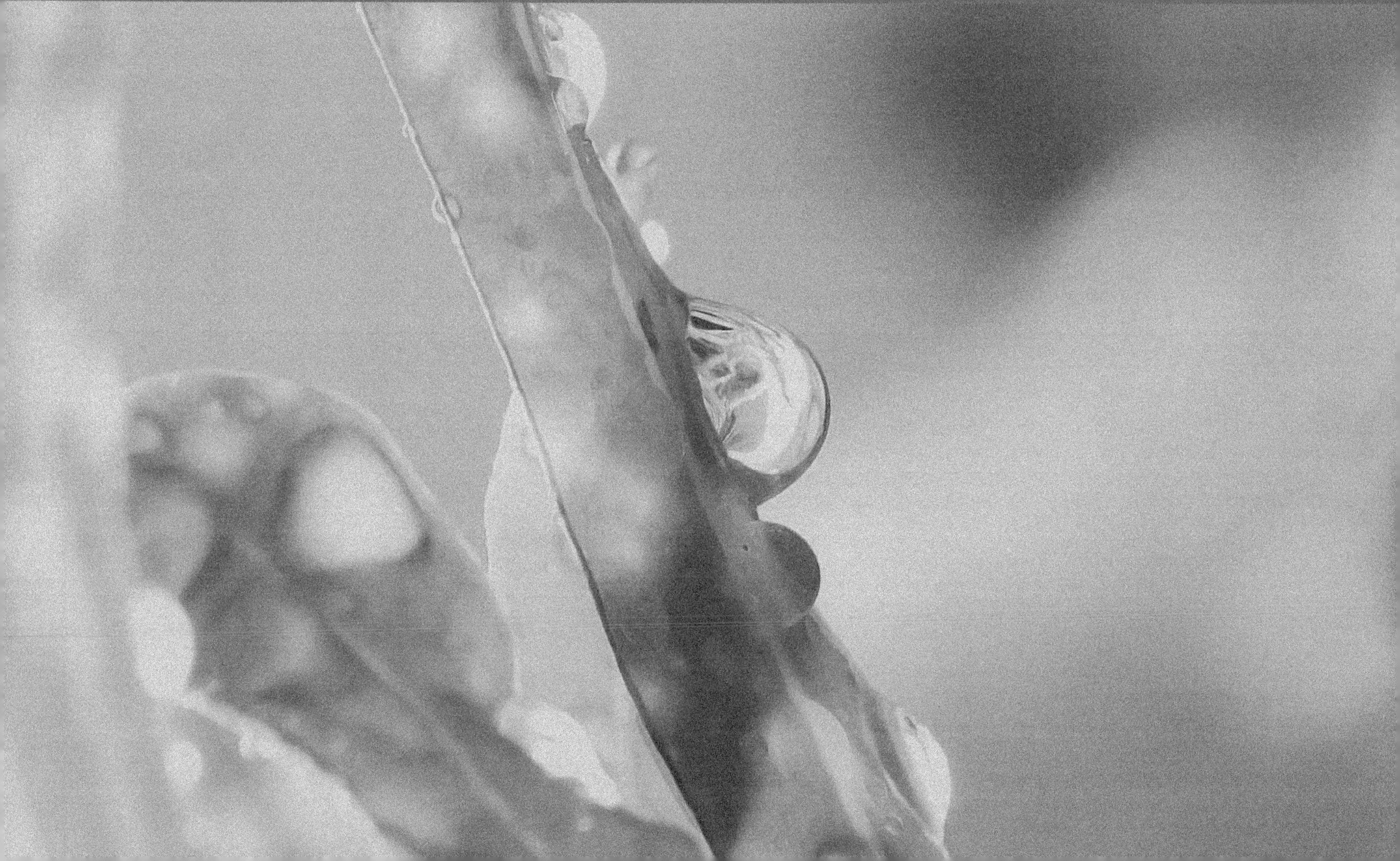

Hydrosphere

Topics include:

- Hydrological cycle within a drainage basin
- Interpretation of hydrographs

The water cycle, also known as the hydrological cycle, refers to the continuous movement of water on, above and below the Earth's surface. Drainage basins are a component of the hydrological cycle. You should be able to draw a diagram to show the hydrological cycle and explain the main elements within the diagram. You should also be familiar with hydrographs and be able to explain and analyse them.

Hydrological cycle

Figure 3.1: Hydrological cycle

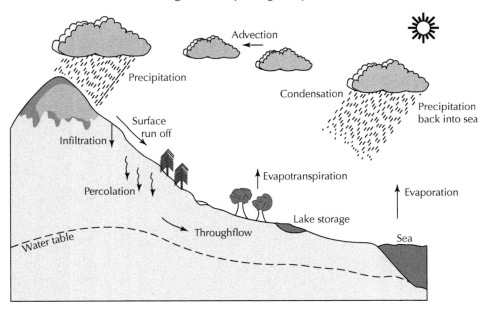

In your exam, you may be given a diagram of the hydrological cycle or asked to draw this yourself. Either way you should be familiar with the key components of the cycle.

Figure 3.2 summarises how the hydrological cycle works.

Figure 3.2: Key components of the hydrological cycle

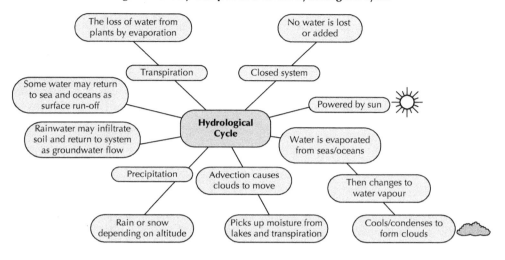

> ## Top Tip
>
> Make sure you are familiar with the main elements of the hydrological cycle. Questions on this topic will expect you to define these elements in relation to the question.

Q Study Figure 3.3. **Explain** how human activities such as those shown in Figure 3.3 can impact on the hydrological cycle. **6 marks**

Figure 3.3: Human activities which affect the hydrological cycle

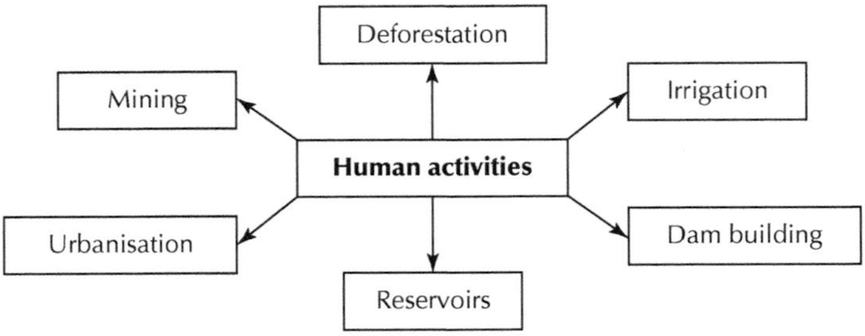

Look at the two answers below. What differences do you notice between them? Think about structure, attention to detail and relevance to question.

Answer A

Deforestation increases the amount of run-off water and it decreases evapotranspiration, therefore decreasing cloud formation. **(1 mark)** *As there are no longer tree roots to take up ground water or leaves to intercept precipitation, this leads to more extreme river flows.* **(1 mark)**

Irrigation takes water from underground water stores therefore lowering the water table. **(1 mark)** *It also places more water in surface stores, i.e. canals/ditches, which will increase evaporation and evapotranspiration.* **(1 mark)** *Furthermore, given irrigation improves plant growth, water will be removed from the hydrological cycle as crops grow.* **(1 mark)**

Urbanisation will reduce the amount of natural vegetation available and replace it with impermeable surfaces and drains. This will increase surface run-off and also evaporation, leading to higher river levels. **(1 mark)** *It will also reduce the water table as less water is being returned to underground stores.* **(1 mark)**

Mining will cause the silting up of lakes, reservoirs and rivers reducing the amount of storage in these areas. **(1 mark)** *It may also result in a reduced vegetation cover, leading to increased run-off and evapotranspiration.* **(1 mark)**

By building dams and reservoirs, water will be stored and therefore less water will flow over the surface and through underground stores, back to the river. This can result in a lowering of the water table. **(1 mark)** *They can also have an impact on where it rains. Dams and reservoirs increase evaporation, resulting in increased cloud formation in these areas. Cloud formation would therefore be altered, changing levels of precipitation.* **(1 mark)**

6 out of 6 marks

Top Tip

For full marks to be awarded, at least 2 factors from the diagram should be referred to.

Answer B

Deforestation affects it as the trees take in the water and release it slowly into the sky and into the land to the flood plain. However if you remove the trees the water will just run down the hillside damaging the soil and affecting the water cycle as the water will return to the river and sea quickly, increasing river flow. (**1 mark**)

Reservoirs and dams affect it too as they stop the water from running down streams into seas and oceans meaning that the water is 'held up', preventing it from moving within the cycle. (**1 mark**)

Urbanisation affects it too as it leads to the building of lanes and streets. This means the trees are removed and that affects it. People in the urbanised area will also take water and use it – slowing down the process.

2 out of 6 marks

What answer comes out on top?

Answer A is the stronger of the two, achieving six out of six, whereas answer B only managed two out of six.

Answer A takes each activity in turn and explains how it can impact on the hydrological cycle. The structure is clear and logical, helping the candidate to regularly link back to the question. Throughout the answer the candidate is specific about what parts of the cycle are affected and how each human activity affects the movement of water in different ways.

Although answer B is structured in a similar way, the candidate fails to answer the question fully with little reference to the hydrological cycle itself. One mark was awarded in the first paragraph for a limited explanation of two points. However, vague statements like *'people in the urbanised area will also take water and use it – slowing down the process'*, add little value to the answer. Here, the candidate should refer to specific areas within the cycle that are affected by specific human activities.

Hydrographs

Figure 3.4: Flood hydrograph

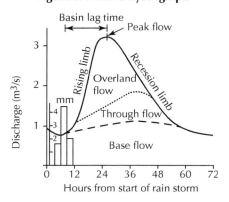

A hydrograph indicates how a river responds to a period of rainfall. Figure 3.6 and 3.7 provide a summary of the key differences between an urban and rural hydrograph.

Figure 3.5: Features of a hydrograph

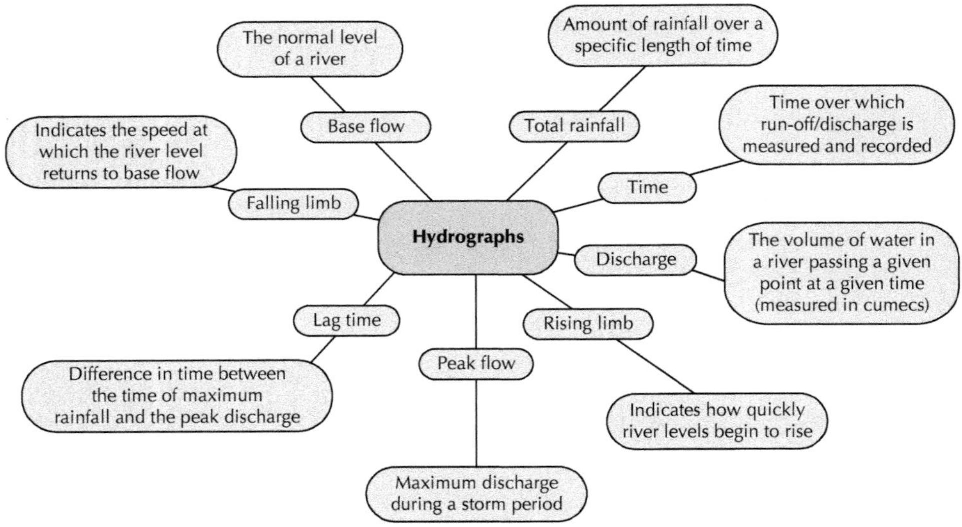

The normal level of a river

Amount of rainfall over a specific length of time

Indicates the speed at which the river level returns to base flow

Base flow

Total rainfall

Time over which run-off/discharge is measured and recorded

Falling limb

Time

Hydrographs

Discharge

The volume of water in a river passing a given point at a given time (measured in cumecs)

Lag time

Rising limb

Peak flow

Difference in time between the time of maximum rainfall and the peak discharge

Indicates how quickly river levels begin to rise

Maximum discharge during a storm period

It is these features that you should refer to if asked to explain a hydrograph or asked to compare hydrographs for two different river basins, i.e. that of an urban and a rural area.

Below you will find a summary of the key differences between these two different river basins.

Figure 3.6: Characteristics of an urban hydrograph

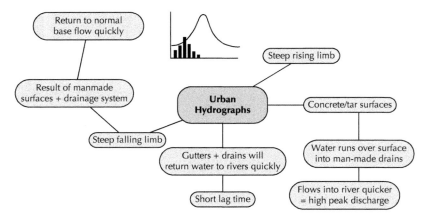

Return to normal base flow quickly

Steep rising limb

Result of manmade surfaces + drainage system

Urban Hydrographs

Concrete/tar surfaces

Steep falling limb

Gutters + drains will return water to rivers quickly

Water runs over surface into man-made drains

Short lag time

Flows into river quicker = high peak discharge

Figure 3.7: Characteristics of a rural hydrograph

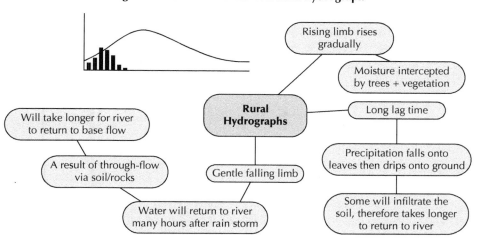

Q Study Figure 3.8. **Explain** the changing river levels on the River Nene in Northampton from 29th to 30th April 2012. **5 marks**

Figure 3.8: Flood hydrograph for the River Nene in Northampton

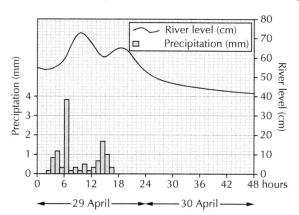

Answer A

The river levels are changing from 29th to 30th April because on the 29th there is 0.2mm of precipitation so not very much is falling so no water can run back into rivers or streams. As the day goes on, at 4 hours the rainfall increases to 0.8mm and continues to rise. This causes the rising limb to rise as there is increased discharge. **(1 mark)**

The peak rainfall was during the 6th hour on the 29th April at 3.8mm. The peak river level was at the 9th hour on the 29th April at 73cm. The river level then decreases to 61cm after 15 hours. The river rises from 61 to a peak of 65cm after 19 hours, then continues to fall to its lowest level of 42 cms.

1 out of 5 marks

What makes this a weak answer?

This candidate has described the graph well. However, they have given little explanation for the changing river levels, therefore failed to answer the question. They have commented on changes to precipitation and linked this to the river level, earning one mark for a limited explanation of two factors. However, the candidate has failed to make any other connections throughout the answer, costing them marks.

Now look at answer B below. Notice how this candidate has structured their answer, regularly linking back to the question.

Answer B

As the rain starts, the river rises quickly as shown by the steep rising limb. This implies the hydrograph shows an urban area. **(1 mark)** *Some rain will* **infiltrate** *the soil. However, due to drains and impermeable road surfaces, most water is directed straight back to the river.* **(1 mark)**

The river reaches its peak of 73cms between 9.00 and 10.00 hours. As the rainfall gets heavier, underground stores become full. **(1 mark)** *The soil becomes* **saturated** *causing water to run off the land and enter the river quicker. This could cause flooding to occur.* **(1 mark)** *Between 18:00 and 48:00 hours there is no rainfall. This causes the river level to decrease gradually to 42cms. From the graph, you can see the falling limb is much more gentle than the rising limbs. This is because water is being added into the river from through-flow almost as quickly as the river is draining it away.* **(1 mark)**

5 out of 5 marks

What makes this a strong answer?

From the outset this candidate makes a connection between increased precipitation and increased river levels. They also correctly identify that this is a hydrograph for an urban area and explain how changes in rainfall affect the flow of the river. This answer is well structured, taking each section of the graph in turn, enabling the candidate to regularly link back to the question.

Top Tip

Be careful when extracting information from a hydrograph. Read the graph carefully and use the correct units.

Maximising Marks

Questions that refer to hydrographs will expect you to extract information before explaining what it means in relation to the question, e.g.

'From the graph you can see the falling limb is much more gentle than the rising limb (reference to graph). *This is because water is being added into the river from through-flow almost as quickly as the river is draining it away* (explanation).' **(1 mark)**

If you simply just describe the change in precipitation or river levels, no marks will be awarded.

Glossary

Advection: The horizontal transfer of a quality such as heat, cold or humidity from one region to another.

Condensation: The process of water changing from a vapour to a liquid. It takes place when water vapour is cooled. It then condenses to form clouds.

Evaporation: The process by which liquid water is transformed into a gas.

Evapotranspiration: The process by which moisture is returned to the atmosphere by direct means through transpiration from vegetation and evaporation.

Infiltration: The movement of water into soil.

Percolation: The downward movement of water through the soil.

Precipitation: Any rain, hail, sleet or snow that falls within the hydrological cycle, depending on the altitude.

Run-off: This is the rainwater that flows over the surface back to streams and river channels.

Through-flow: The downslope movement of water through the lower soil towards streams or rivers.

Transpiration: The loss of water from leaves or stems by evaporation.

Water table: The boundary between water saturated ground and unsaturated ground.

Section 4: Lithosphere

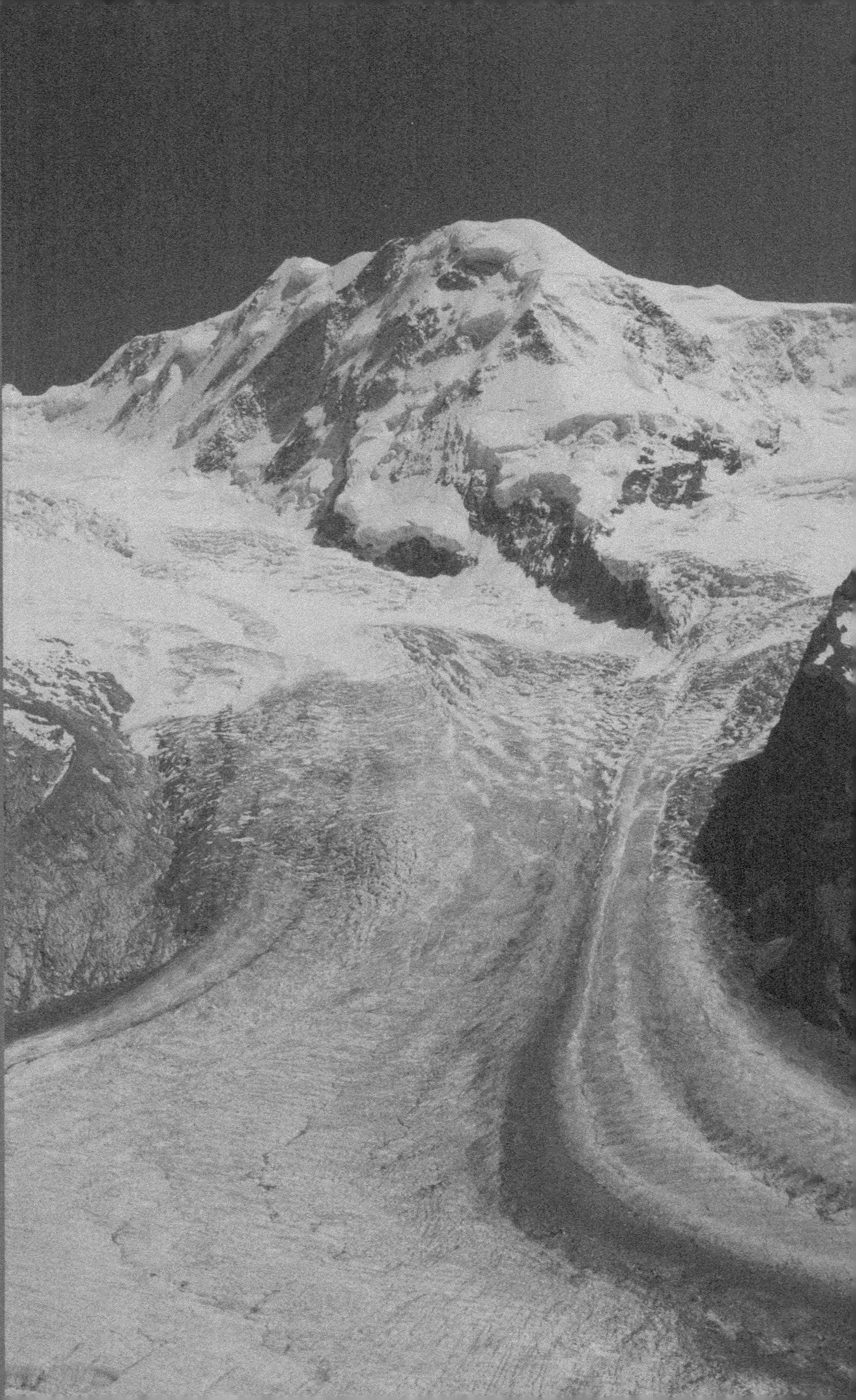

Lithosphere

Topics include:

- Formation of erosional and depositional features in glaciated and coastal landscapes
- Rural land use conflicts and their management related to glaciated and coastal landscapes

You should be familiar with erosional and depositional features in both a glaciated and coastal landscape. You should be able to explain the processes involved and draw simple line diagrams to illustrate how these features are formed. You should also be able to explain the conflicts that arise in these areas and discuss the management strategies employed to minimise their impact.

Glaciated landscapes

Erosional features

You should be able to explain the formation of the following erosional features and draw annotated diagrams where appropriate:

- Corrie
- Arete
- Pyramidal peak
- U-shaped valley
- Hanging valley

Q **Explain** the conditions and processes involved in the formation of a u-shaped valley. You may use diagrams in your answer. **5 marks**

Answer A

A u-shaped valley forms when a glacier moves down through an existing v-shaped valley, eroding the land to create a u-shape. **(1 mark)** *Plucking, abrasion and frost shattering help to erode the valley.* **(1 mark)** *The ice steepens, deepens and widens the valley. A misfit stream is often left behind.* **(1 mark)**

Pupil diagram A: U-shaped valley with no annotation

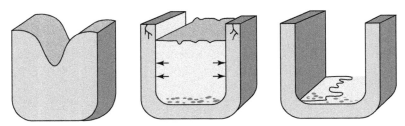

3 out of 5 marks

What makes this a weak answer?

Answer A is largely descriptive with few developed points. Although the candidate has mentioned the processes involved they have not explained directly how they contribute to the formation of this feature. The candidate has included three well-structured diagrams. However, without annotations they cannot achieve any marks. As a result of poor exam technique, opportunities have been lost, leaving this candidate with a total of 3 out of 5.

Now compare this with the answer below. What differences do you notice?

Answer B

A glacier flows down an existing v-shaped valley. Interlocking spurs are removed and truncated spurs are left behind, as the glacier 'bulldozes' its way downhill. **(1 mark)**

The immense weight of the glacier as it moves downhill aids processes such as plucking and abrasion. **(1 mark)** *Plucking occurs when ice freezes onto the sides of the valley and rips off chunks of rock as it moves, steepening the valley sides.* **(1 mark)** *These rocks will assist abrasion – where rocks trapped at the bottom of the glacier scour away the landscape, deepening the valley.* **(1 mark)** *Freeze-thaw action further assists the process.*

As the ice melts, a u-shaped valley is left behind with steep sides and a flat valley floor. A misfit stream or ribbon lake may be found on the valley floor. **(1 mark)**

Pupil diagram B: U-shaped valley with annotation

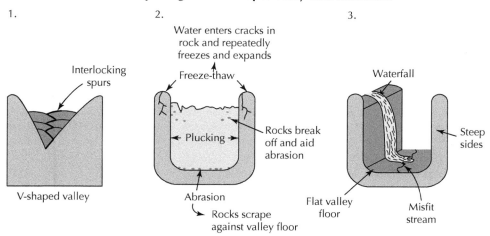

1. Interlocking spurs

V-shaped valley

2. Water enters cracks in rock and repeatedly freezes and expands

Freeze-thaw

Plucking

Rocks break off and aid abrasion

Abrasion

Rocks scrape against valley floor

3. Waterfall

Steep sides

Flat valley floor

Misfit stream

5 out of 5 marks

What makes this a strong answer?

This is a well-structured answer where the candidate has explained clearly the conditions **and** processes involved in the formation of a u-shaped valley. Although freeze-thaw action is not fully developed within the written answer, the candidate explains this well through use of an annotated diagram. Notice how the candidate has kept the diagrams simple yet used them effectively through the addition of annotations. This is a good example of how annotated diagrams can be used to help strengthen an answer and would have helped the candidate gain additional marks if needed.

Depositional features

You should be able to explain the formation of the following depositional features and draw annotated diagrams where appropriate:

- Drumlins
- Eskers
- Terminal moraine
- Outwash plain

Q **Explain** the formation of a drumlin. You may use annotated diagrams in your answer.
5 marks

Look at the two answers below. What differences do you notice between them? Think about structure, use of diagrams and attention to detail.

Answer A

Drumlins are elongated hills made up of glacial deposits. They are made up of unsorted material because the ice deposited all the material at the same time underneath the glacier as it moved downhill. **(1 mark)** *The steep slope faces up-valley whilst the more gently sloping 'lee' slope faces down-valley. Drumlins are formed when the sediment becomes too heavy for the glacier. The glacier deposits the material, shaping it into streamlined mounds as it flows over the top.* **(1 mark)** *If there is a small obstacle on the ground, this may act as a trigger point and material can build up around it.* **(1 mark)**

Drumlins may also form when the ice flows over a layer of sediment that is deeper than normal. The huge overlying weight of ice causes increased pressure on the sediment beneath it, moulding it into the drumlin shapes. **(1 mark)** *Further ice movement can reshape drumlins after they have been originally deposited. Drumlins are found in swarms or in a 'basket of eggs' topography.* **(1 mark)**

Pupil diagram A: Formation of a drumlin

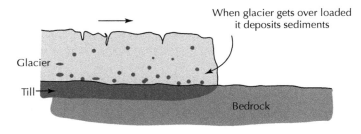

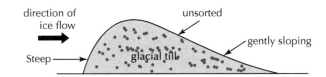

5 out of 5 marks

Answer B

Drumlins are streamlined mounds of material. They formed after the ice age when temperatures were higher and material was dropped. They often form together in groups and look like eggs on their side. They are formed when a glacier is moving downhill and the material becomes too heavy for the ice to carry. **(1 mark)** *The material is then dropped all together in a big mound. As the ice moves over it, the material is reshaped into a sloping hill. The end points in the direction in which the ice was moving.* **(1 mark)** *Glasgow is built on a swarm of drumlins.*

2 out of 5 marks

What answer comes out on top?

Answer A is a well-structured, detailed response. The candidate not only explains how drumlins are formed but provides an alternative method to their formation. The inclusion of annotated diagrams helps to strengthen this answer and would have earned marks if full marks had not already been awarded.

In comparison, answer B is much more basic, achieving just 2 out of 5. Although the candidate clearly knows what a drumlin is, they do not make it clear how drumlins form. In order for marks to be awarded you need to make it clear to the examiner how different conditions allow material to be deposited to allow drumlins to develop. The inclusion of a simple diagram may have helped this candidate gain additional marks.

Annotated diagrams

Below are some examples of the simple line diagrams you should be familiar with for glaciated landscapes. If asked to explain the formation of any of these features remember to explain the conditions and processes in detail. Using diagrams will help you to do this and can save you time in the exam.

Figure 4.1: Formation of a corrie

1. Before glaciation

North-facing

Snow compacts = ice

2. During glaciation

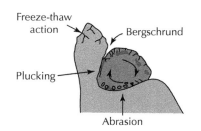

Freeze-thaw action

Bergschrund

Plucking

Abrasion

3. After glaciation

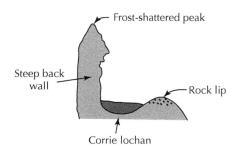

Frost-shattered peak

Steep back wall

Rock lip

Corrie lochan

Figure 4.2: Formation of an arête

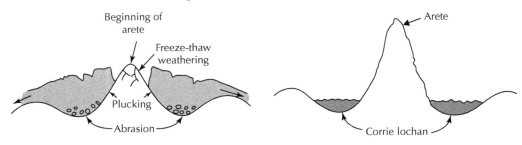

Figure 4.3: Formation of a pyramidal peak

Figure 4.4: Formation of a u-shaped valley

1. Before Glaciation

2. During Glaciation

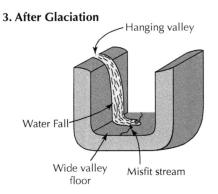

3. After Glaciation

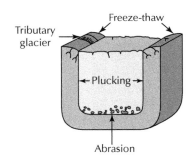

Top Tip

In a question which states, *'With the aid of annotated diagrams ...'* marks will be lost if diagrams are not included. However, if the question gives you the option, for example, *'you may use annotated diagrams'*, diagrams do not need to be included for full marks to be awarded.

Coastal landscapes

You should be able to explain the formation of the following coastal features and draw annotated diagrams where appropriate:

- Headlands and bays
- Caves
- Arches
- Stacks
- Stumps

Below is a step-by-step guide to explaining the formation of caves, arches, stacks and stumps.

Figure 4.5: Caves, arches, stacks and stumps

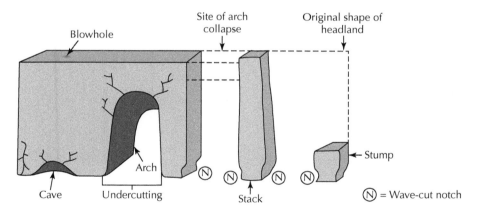

1. Draw the diagram first. As you write your answer, work your way along the diagram. This will not only help you structure your answer but will help ensure you don't miss out vital information.

2. Start with a **joint/line of weakness** in the rock. Think about what processes are involved to turn this joint into a **cave**? You should be able to explain hydraulic action, attrition, solution and abrasion.

3. At times, a **blowhole** may form. Where does this happen? What causes this feature to form?

4. What then has to happen to the cave for an **arch** to form?

5. Once an arch has formed, what processes are involved to turn the arch into a stack? What happens to the roof of the arch?

6. Once a stack is left isolated from the headland, what processes are involved to turn this stack into a stump?

When explaining the formation of physical features, try to think logically about the processes involved. Make sure you have explained each stage fully, allowing you to move on.

Coastal deposition

Figure 4.6 illustrates the features of coastal deposition that you should be familiar with.

Figure 4.6: Coastal deposition landscape

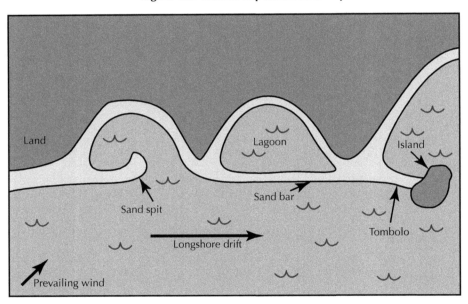

When explaining the formation of any of these features you should be able to explain the process of **longshore drift**. When explaining longshore drift it is useful to include a quick diagram. This will help make your answer clearer and, if done well, will save you time in the exam.

Figure 4.7: Longshore drift

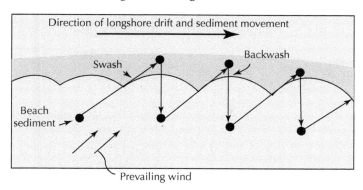

Q Explain how **one** of the following features of coastal deposition is formed. You may use diagrams in your answer.

- Spit
- Beach
- Tombolo

4 marks

Look at the answers below. What differences do you notice between the two answers? Think about structure, relevance to the question and attention to detail.

Answer A

Bays are formed where there are layers of hard rock and soft rock. The soft rock is eroded away quickly and headlands are left where the hard rock is left sticking out to sea. Bays are left in between two headlands. Bays are more sheltered and allow a range of features to form. Longshore drift is a defining factor of depositional features. A bar, spit and tombolo are all common features of coastal deposition. They are formed by material being deposited in a narrow ridge. However, depending on what the ridge joins on to will determine what name is given to the feature.

0 out of 4 marks

Answer B

A spit

A spit is a long, narrow ridge of sand or shingle that sticks out into the sea and is joined at one end to a headland. **(1 mark)** *Longshore drift is the main process involved in its formation. Longshore drift is the gradual movement of material along a beach caused by the waves hitting the beach at an angle (see diagram).* **(1 mark)** *As the spit builds out to sea it becomes increasingly affected by the waves and the wind. This causes the end to curve in towards the shore, creating a hook-like shape.* **(1 mark)** *The area of standing water behind a spit is sheltered from the harsh winds and waves and is therefore calmer. This allows salt marshes to form.* **(1 mark)**

Pupil diagram B: Pupil diagram of longshore drift

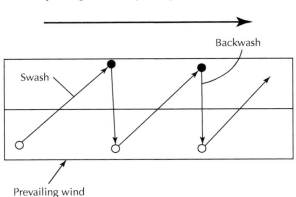

4 out of 4 marks

What answer comes out on top?

Answer A is vague and bears little resemblance to the question. The question specifically asks for the formation of **one** feature from a given list. However, the candidate starts by describing erosional features then gives a list of depositional features. Although, the candidate makes reference to longshore drift they do not explain what it is or how it contributes to the formation of any one feature, resulting in opportunities being lost. This is a very weak answer reflecting a candidate who has dived into the question. As a result, no marks have been awarded.

In comparison, answer B is well structured, making it clear from the beginning what feature they have chosen. Stating your choice at the beginning of the answer not only makes it clear to the examiner what feature you are explaining but also helps to keep you focused as you work your way through the answer. The candidate has explained the process of longshore drift and further develops this through the use of a diagram. The diagram is a useful addition to this answer and would have earned the candidate additional marks if they were needed. Overall, this response is logical and detailed, earning the candidate full marks.

Physical features on a map

You should be able to identify glaciated and coastal features on an OS map. Figures 4.8 and 4.9 illustrate how some of these features can be identified.

Glaciated features

Figure 4.8a: A hanging valley

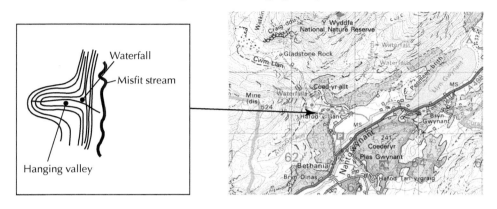

Figure 4.8b: A u-shaped valley

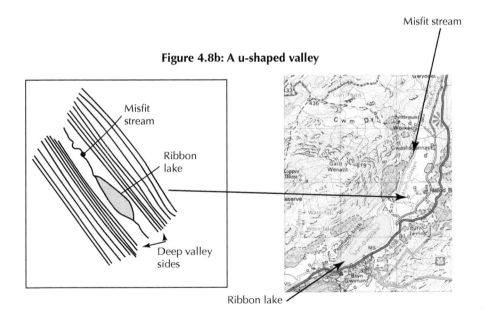

Figure 4.8c: A corrie

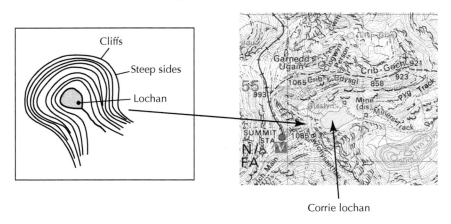

Corrie lochan

Figure 4.8d: An arête

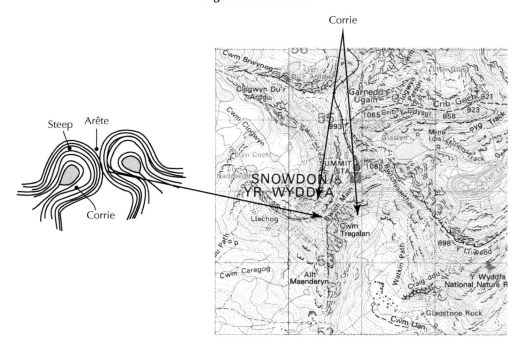

Figure 4.8e: Pyramidal peak

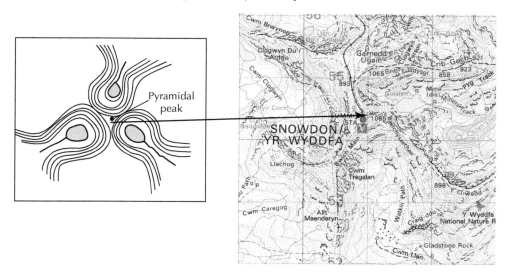

Coastal features

Figure 4.9a: Shingle beach, sandy beach and sand dunes

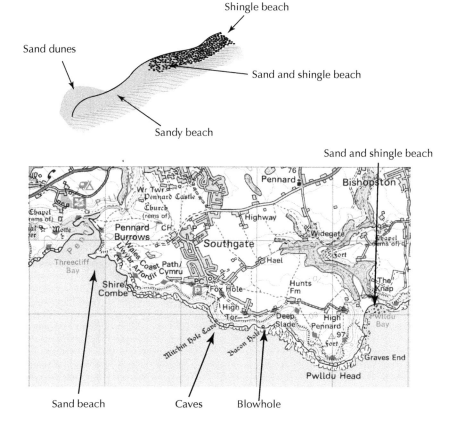

Figure 4.9b: Headland, stack and stump

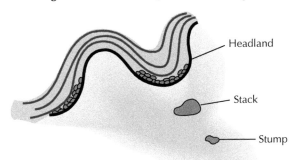

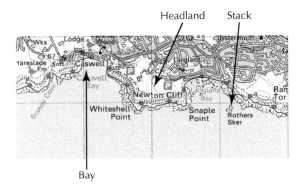

Figure 4.9c: Cliffs and wave-cut platform

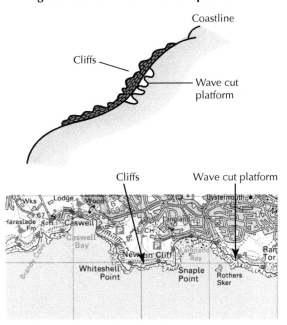

Figure 4.9d: Spit, bay and tombolo

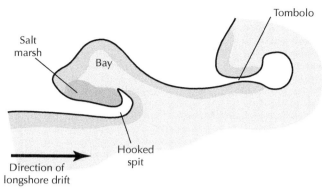

Maximising Marks

Maximising marks through use of diagrams

- Always include diagrams
  - o Even if the question doesn't ask for them, you should include diagrams. They may help you pick up an extra mark and can save you time in the exam.
- Structure
  - o Splitting your diagrams into different stages can help structure your answer, e.g. before, during and after glaciation. This may also help to ensure you don't miss anything out in a hurry.
- Keep them simple
  - o No marks are awarded for your artistic ability so keep them clear and simple to avoid wasting time.
- Include annotations
  - o Diagrams become meaningless if they are not annotated. Don't waste time by leaving your drawings blank.
- Avoid re-writing your answer
  - o Use your annotations wisely. They should add to your answer without simply repeating what you have already said.

Rural land use conflicts and their management

As well as having to explain how physical landscapes are formed, you should also be familiar with how these landscapes are used and the conflicts that occur between different land users. You should be able to explain these conflicts and discuss how effectively they are managed.

Land users that often conflict with each other include:

- Tourists vs farmers
- Tourists vs locals
- Active tourists vs passive tourists
- Quarrying vs farmers
- Quarrying vs tourists
- Conservationists vs tourists

> **Q** For Studland or any other named coastal area you have studied:
>
> **(a)** Explain the environmental problems and conflicts that may arise from competing land uses.
> **6 marks**
>
> **(b)** Discuss the measures taken to resolve environmental conflicts related to tourism.
> **5 marks**

Answer A

(a) *There can be over 100,000 people visiting this area each year, with 36% visiting during the summer months and over* **90%** *of these arriving by car. This causes conflict with the local community. The high number of tourists arriving by car means that traffic congestion is a big problem.* **(1 mark)** *The car parks at Studland and Lulworth Cove have limited access so there is a concentration of cars in this area, increasing air and noise pollution.* **(1 mark)** *An increase in tourists can also bring an increase in holiday home ownership. This can cause house prices to increase, resulting in rural depopulation, which leaves rural areas empty during the week or off-peak periods.* **(1 mark)**

The Ministry of Defence uses land behind Lulworth and Studland for training purposes as well as barracks. This causes conflicts with the tourists, especially walkers, as the MoD closes roads and coastal footpaths during exercises. **(1 mark)** *This could upset tourists as it is the coastline that is Dorset's main attraction and access to some beaches may only be possible at weekends.* **(1 mark)**

There are many important conservation areas here. These include an RSPB reserve, an Area of Outstanding Natural Beauty, Site of Special Scientific Interest and a Heritage Coastline. Conservationists feel that the large numbers of tourists using the area, especially during the summer months, are causing environmental damage. **(1 mark)** The coastal footpaths are being deeply eroded, creating eyesores on the coastline. **(1 mark)** Prevention methods can result in access being limited in some places, causing conflict with tourists. **(1 mark)**

Farmers also come into conflict with tourists as they walk across farmland, leave gates open and restrict access to fields by irresponsible parking. **(1 mark)** Litter dropped by tourists can be swallowed by animals, causing them to choke. **(1 mark)**

6 out of 6 marks

(NB: 1 mark can be awarded for 2 specific named examples)

(b) Around the **Studland** area the four main car parks have been expanded and can now accommodate another **820 cars** in an attempt to stop cars parking on grass verges and other inappropriate parking. **(1 mark)** Park and Ride schemes have also been introduced. However, due to the sheer number of visitors arriving by car, this is a difficult problem to solve. **(1 mark)**

Some paths have been closed to prevent further erosion and sand dunes have been fenced off. The fences collect sand where trampling or wind erosion has taken hold. **(1 mark)** Boardwalks are laid along main footpaths to reduce footpath erosion. To reduce litter, bins are put along the paths and at the back of beaches. **(1 mark)** Fires are started by irresponsible tourists so fire beaters are available and in some areas fire breaks have been formed. **(1 mark)**

The public are being educated on how to look after the environment through the use of leaflets and information boards. Rangers are also available to give advice to the public on protection of the environment but this is expensive and not always adhered to. **(1 mark)** Organisations like the National Trust and English Nature have taken on some of the responsibility for looking after the coastline. The money charged by the National Trust for the facilities provided is used to further protect the area and also for improvements to the beach and its facilities. **(1 mark)**

5 out of 5 marks

What makes this a good answer?

Part (a) is very well structured, with the candidate taking a new paragraph for each conflict. A variety of conflicts have been included, with each one being explained in detail. It is good to include case study knowledge. However, this candidate has repeated the use of named examples, which adds little value to the answer and wastes time.

Part (b) has been developed well and regularly links back to the question. The candidate has discussed each measure fully with full marks being well deserved.

Answer B

(a) *With over 100,000 people visiting each year, the World Heritage Coastline of Dorset is used by a wide variety of people. Large numbers of people want to use the coastline for different activities, which results in conflict.*

At any one time, there are approximately 18 different activities happening at Poole Harbour (example). Poole Harbour is home to the main ferry route across the channel to Cherbourg. Many tourists also come here to do water sports.

Tourists can also conflict with locals by causing traffic congestion at peak times. Studland Heath (example) attracts ~20,000 visitors per day with 90% of people arriving by car. These tourists may also bring their dogs and allow them to 'do their business' on the pavement, which leaves the place looking untidy. **(1 mark)**

Approximately 200,000 people per year walk the 3km coastal walk between Lulworth Cove and Durdle Door. This has resulted in footpath erosion as large numbers of people put pressure on the path, causing it to wear away. **(1 mark)** *Rural depopulation is also occurring and so the government is trying different schemes to try and solve the issue.*

3 out of 6 marks

(b) *The authorities have put a number of measures in place to try and resolve these conflicts. Some strategies have been more effective than others, these include:*

- *Zoning*
- *Fencing off footpaths*
- *Park and ride schemes*
- *A 'no-firing' policy at weekends*

These strategies can be effective to an extent. However, putting restrictions on tourists may make them unhappy and they will just go elsewhere.

0 out of 5 marks

What makes this a weak answer?

Through the inclusion of specific named examples, facts and figures in part (a), it is clear this candidate has good case-study knowledge. The inclusion of at least 2 specific named examples earns the candidate 1 mark. However, the issue with this answer lies with the lack of explanation. In most cases, the candidate has stated what the problem is and given examples without going on to explain the conflict in detail. Given this answer is largely descriptive, the candidate can receive a maximum of 3 out of 6.

As for part (b), the first paragraph is vague earning the candidate no marks. The candidate then proceeds to state the solutions in a list format, which has proved detrimental to the overall number of marks awarded. The last sentence is a very valid point. However, the candidate has not discussed what these restrictions are and has therefore failed to answer the question. This answer has achieved 0 out of 5.

Glossary

Abrasion: Rocks stuck in the bottom of the glacier grind away the bedrock (like sandpaper) under the glacier.

Alluvial fan: Fan-shaped deposits of material transported by water *(alluvium)*.

Arete: A knife-edged ridge between two corries.

Attrition: As rocks are carried by water, they smash together and break into smaller, smoother and rounder particles.

Corrie: An armchair-shaped hollow.

Drumlin: An elongated mound of unsorted glacial till formed by the streamlined movement of glacial ice.

Erratic: A rock/boulder transported by a glacier to an area of different geology.

Esker: A long, winding ridge of stratified sand and gravel deposited by meltwater streams.

Freeze-thaw action (Frost-shattering): Water enters into cracks in the rock. As temperatures drop, the water freezes and expands. As temperatures rise, the ice melts. As this process repeats over time, the rock gradually weakens, forcing fragments to break off.

Hanging valley: A small valley left 'hanging' above a large u-shaped valley formed after a tributary glacier could not erode as far down as the glacier in the main valley.

Hydraulic action: Erosion caused by the sheer force of water breaking off material from the coastline.

Misfit stream: A stream that no longer fits the valley floor.

Moraine: Material deposited by a glacier.

Plucking: Ice freezes into cracks in rocks and when the glacier moves, it rips out chunks to leave a jagged surface.

Till: Unsorted material deposited by glacial ice.

U-shaped valley: a valley with steep sides and a flat valley floor, carved out by a glacier.

Weathering: The breakdown of rocks through physical and chemical processes caused by the weather.

Section 5: Biosphere

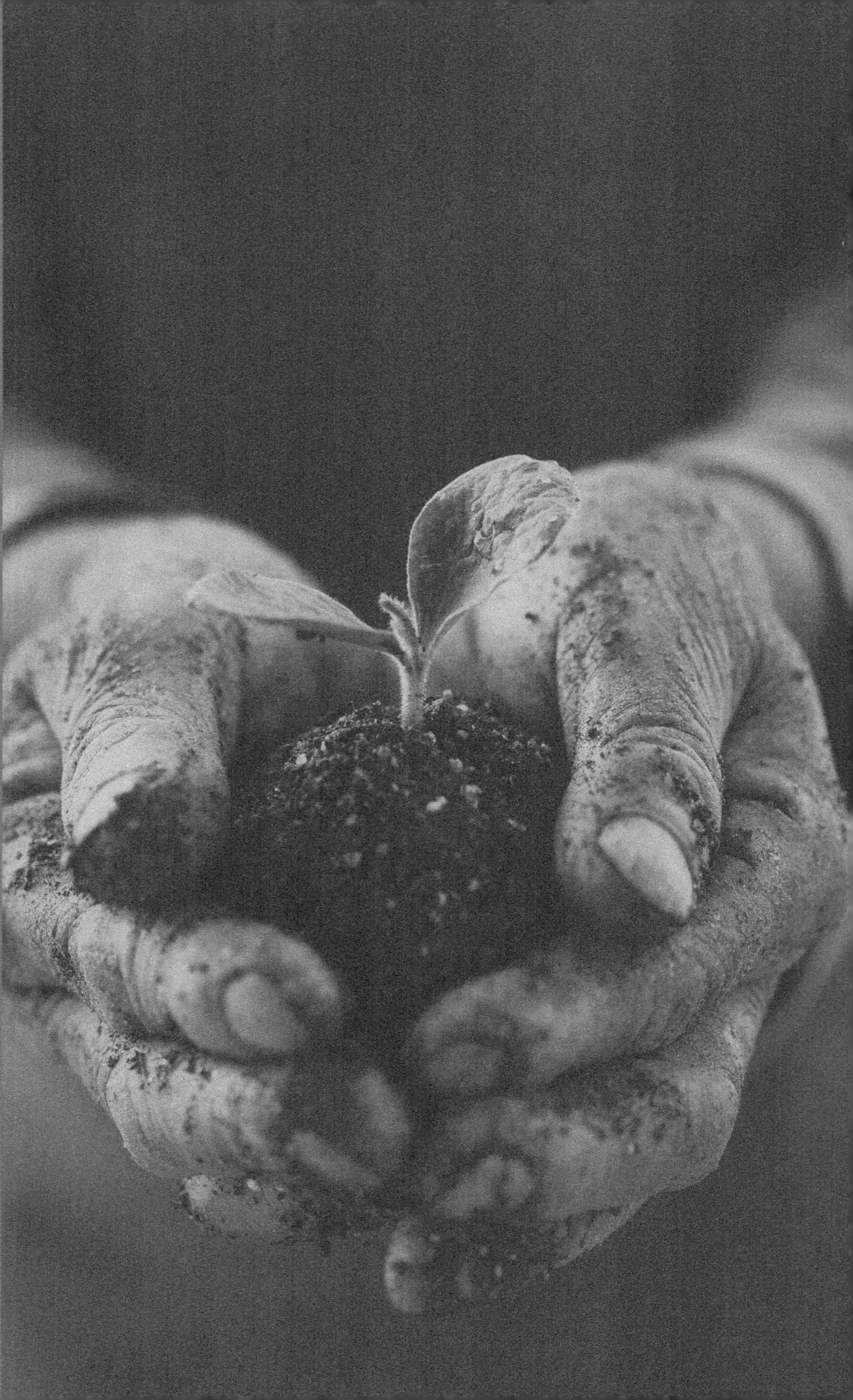

Biosphere

Topics include:

- The properties of podzol, brown earth and gley soils
- The formation processes of podzol, brown earth and gley soils

These questions will be on the properties and/or formation processes of a podzol, brown earth or gley soil profile. You should be able to explain how a soil profile is formed in relation to a number of factors, namely climate, relief, organisms, parent material and drainage. You should also be able to draw and annotate specific soil profiles.

Q **Explain** how the major soil forming factors such as natural vegetation, soil organisms, climate, relief and drainage have contributed to the formation and characteristics of a brown earth soil.

6 marks

Look at the two answers below. What differences do you notice between them? Think about structure, level of detail and relevance to the question.

Answer A

Natural vegetation affects the formation of a brown earth. The deciduous vegetation provides nutrient-rich litter, which explains why the humus is fertile and mildly acidic. **(1 mark)** *The climate is generally warm and dry – more suitable for soil biota, e.g. earthworms to thrive and mix/aerate the soil. These moving organisms in the soil create non-distinct horizons in brown earth.* **(1 mark)**

The longer roots break up the parent material, making the soil more fertile. Temperature directly affects the rate of biological and chemical activity in the soil, which determines whether mull or mor humus is formed. The B horizon has a lighter brown horizon because of the calcium and

> **Top Tip** ✓
>
> When answering a question like this, you do not need to refer to all factors in the question for full marks. However, reference to at least two factors is recommended.

magnesium that has been leached out. Relief and drainage go hand in hand to influence soil formation. Brown earths are usually found on gently sloping land and hillsides, allowing the soil to be free draining. However, if an iron pan develops due to leaching, free drainage could be prevented. **(1 mark)**

3 out of 6 marks

Answer B

Deciduous forest vegetation provides a deep leaf litter, which, due to the mild climate, is broken down quickly. **(1 mark)** The rapid decomposition of leaf litter creates an A horizon that is nutrient rich. **(1 mark)** The long tree roots penetrate deep into the soil, ensuring the recycling of minerals back up to the surface. **(1 mark)**

Soil organisms such as worms help to vertically mix the soil, aerating it and preventing the formation of clearly defined layers. **(1 mark)** They also help to break down the leaf litter creating a mildly acidic mull humus (pH 5–6.5). **(1 mark)**

Downward leaching occurs as precipitation slightly exceeds evaporation. As minerals move down through the soil, the colour of each horizon varies from black humus to dark brown in A horizon to lighter brown in B horizon where humus content is less obvious. **(1 mark)** Leaching of the most soluble minerals can cause an iron pan to form, which may impede drainage. **(1 mark)**

Brown earths are often found on well drained slopes with little accumulation of excess water. This limits the degree of leaching within the soil. **(1 mark)**

6 out of 6 marks

What answer comes out on top?

Having achieved full marks, answer B is the strongest of the two, with answer A achieving just 3 out of 6.

What makes answer A a 'weak' answer?

If asked to refer to a given soil type your answer should be specific. Vague or irrelevant points will receive no marks. Although this answer contains some relevant information about a brown earth profile, the second paragraph is over-generalised. For example, the candidate talks about the formation of humus yet does not apply this information to the specific soil type. There are also inaccuracies within this answer with the candidate stating the 'B horizon has a lighter brown horizon'. This is a careless mistake and shows a lack of understanding. Therefore, a combination of poorly developed points, a lack of knowledge and inadequate detail has resulted in an answer worthy of just half marks.

What makes answer B a strong answer?

Candidate B on the other hand has used the question to help structure their answer, providing sufficient detail for each soil-forming factor. All information is accurate and relevant, ensuring maximum use of time and each point is well developed, adding depth to their answer. This is an answer typical of an A student.

Soil profiles

Figures 5.1 – 5.3 show the three soil profiles you could be asked about in the final examination.

Figure 5.1: Podzol soil profile

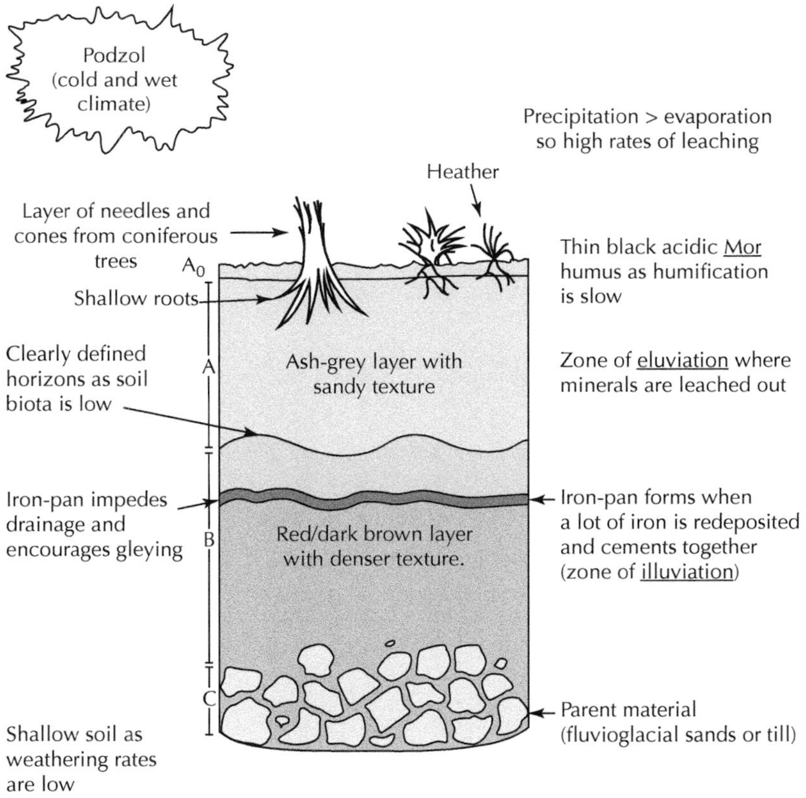

Podzol
(cold and wet climate)

Precipitation > evaporation so high rates of leaching

Heather

Layer of needles and cones from coniferous trees A_0

Shallow roots

Thin black acidic Mor humus as humification is slow

Clearly defined horizons as soil biota is low

A

Ash-grey layer with sandy texture

Zone of eluviation where minerals are leached out

Iron-pan impedes drainage and encourages gleying

B

Red/dark brown layer with denser texture.

Iron-pan forms when a lot of iron is redeposited and cements together (zone of illuviation)

C

Shallow soil as weathering rates are low

Parent material (fluvioglacial sands or till)

Figure 5.2: Brown earth soil profile

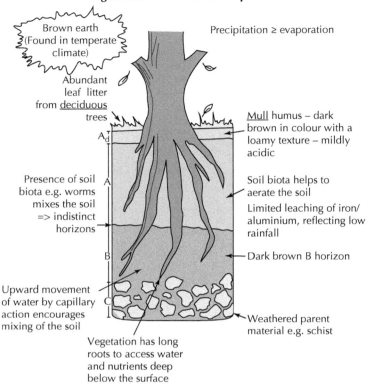

Brown earth (Found in temperate climate)

Precipitation ≥ evaporation

Abundant leaf litter from <u>deciduous</u> trees

Mull humus – dark brown in colour with a loamy texture – mildly acidic

Presence of soil biota e.g. worms mixes the soil => indistinct horizons

Soil biota helps to aerate the soil

Limited leaching of iron/ aluminium, reflecting low rainfall

Dark brown B horizon

Upward movement of water by capillary action encourages mixing of the soil

Weathered parent material e.g. schist

Vegetation has long roots to access water and nutrients deep below the surface

Figure 5.3: Gley soil profile

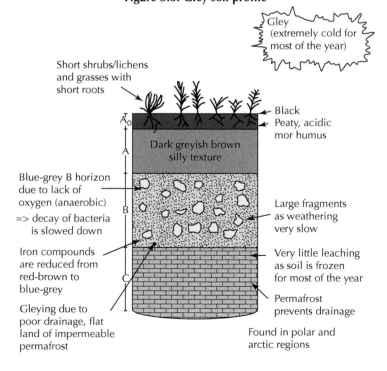

Gley (extremely cold for most of the year)

Short shrubs/lichens and grasses with short roots

Black Peaty, acidic mor humus

Dark greyish brown silly texture

Blue-grey B horizon due to lack of oxygen (anaerobic) => decay of bacteria is slowed down

Large fragments as weathering very slow

Iron compounds are reduced from red-brown to blue-grey

Very little leaching as soil is frozen for most of the year

Gleying due to poor drainage, flat land of impermeable permafrost

Permafrost prevents drainage

Found in polar and arctic regions

If asked to draw and annotate a soil profile, keep the drawing simple. It is the annotations that will earn you marks. Below is a step-by-step guide to annotating a soil profile.

1. Begin by labelling the **different horizons** within the profile before discussing each layer in turn.

Figure 5.4: Standard soil profile

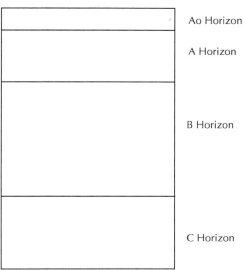

	Ao Horizon
	A Horizon
	B Horizon
	C Horizon

2. Starting with the topsoil, note the thickness of this layer referring to the **type of vegetation, degree of leaf litter** and **depth of humus**. You may also want to make reference to the **roots** – are they short or do they travel all the way down the profile?

3. For the A, B and C horizons, you should be able to explain their:

   (a) Colour

   (b) Texture

   (c) Water content – i.e. free draining/waterlogged?

   (d) You should explain these properties by referring to factors such as climate, vegetation, processes of leaching, eluviation, illuviation and soil biota.

   (e) Are horizons well defined or has mixing occurred?

4. A profile that is purely descriptive will achieve no marks. It is important you refer back to the question and do exactly what it is asking you to do.

Top Tip

If asked to annotate a soil profile you must draw a diagram **then** annotate it fully. An answer that is made up of a simple soil profile **and** a separate paragraph will lose marks.

Soil profile summary

Do not assume that you need only revise one soil type. You may be asked to explain the differences between two soils or even refer to all three soils in an answer. Below is a summary of the key characteristics of each soil type you should be familiar with.

Figure 5.5: Brown earth

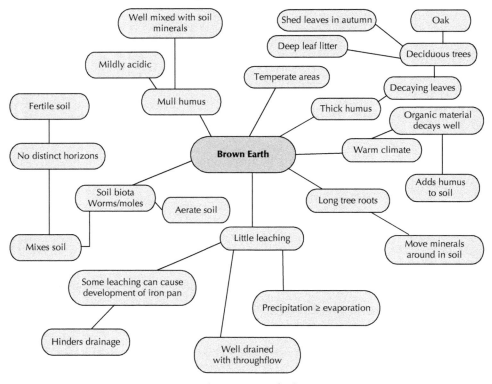

Figure 5.6: Podzol

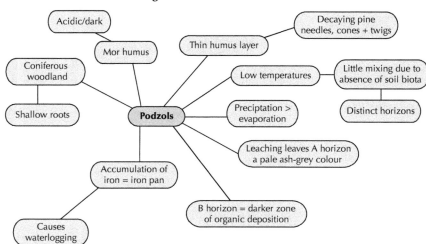

Figure 5.7: Gley

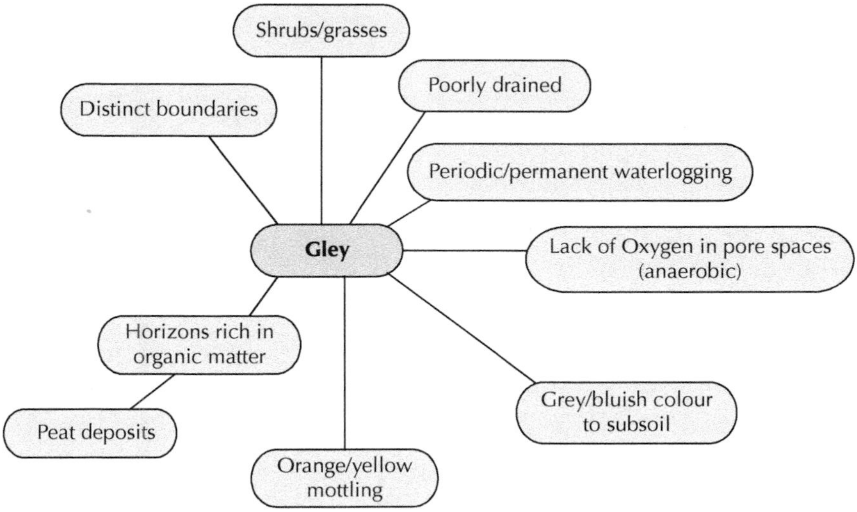

```
                    Shrubs/grasses

   Distinct boundaries        Poorly drained

                          Periodic/permanent waterlogging

            Gley                    Lack of Oxygen in pore spaces
                                            (anaerobic)

   Horizons rich in
   organic matter
                                        Grey/bluish colour
   Peat deposits                            to subsoil

                 Orange/yellow
                    mottling
```

Maximising Marks

For 1 mark, you should give one detailed explanation, e.g.

*'Given the **mild climate**, the litter from deciduous trees and grasses **decomposes quickly**, producing an A horizon that is **rich in nutrients**.'* **(1 mark)**

OR 1 mark can also be achieved through limited description/explanation of **two** factors, e.g.

*'Soil organisms, like worms, break down the leaf litter producing **mull humus**. They also work to mix the soil, ensuring **horizons are not clearly defined**.'* **(1 mark)**

Glossary

Capillary action: The upwards movement of water in a soil.

Eluviation: The transportation of dissolved material within the soil by the movement of water when rainfall exceeds evaporation.

Horizon: A layer parallel with the land surface whose characteristics differ from the layers above and below.

Illuviation: The accumulation of dissolved soil materials as a result of eluviation from one horizon to another.

Leaching: The process through which minerals are carried through the soil by the downward movement of water.

Mull humus: Develops beneath deciduous woodland where leaves are rich in minerals. There is no clearly defined humus layer.

Mor humus: Associated with cooler, wet climates. Worms are not common so there is limited mixing of organic and mineral material. Often found under coniferous woodland.

Parent material: The name given to weathered rock on which soil horizons form.

Regolith: Weathered bedrock.

Section 6: Population

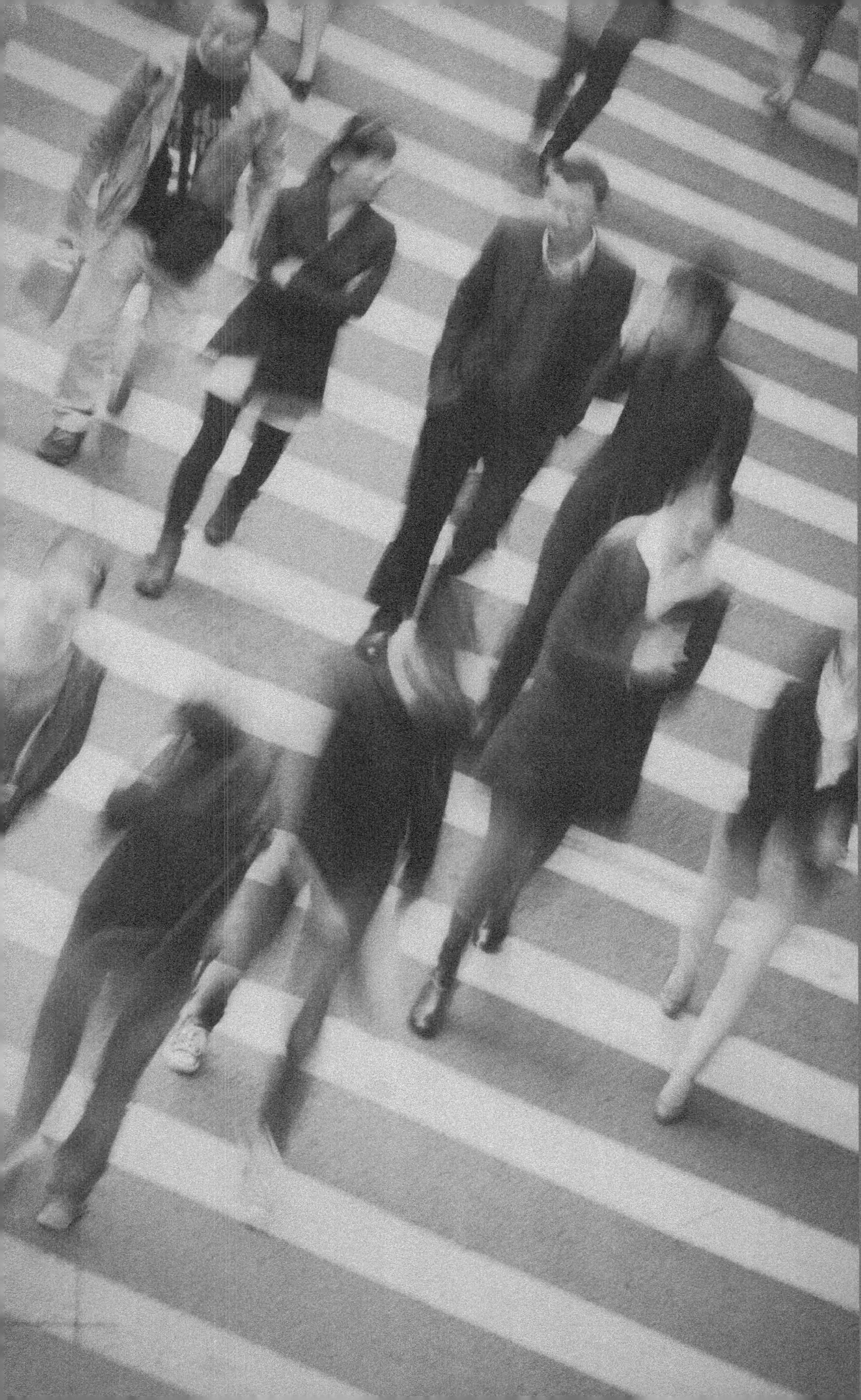

Population

Topics include:

- Methods and problems of data collection
- Consequences of population structure
- Causes and impacts of forced and voluntary migration

Methods and problems of data collection

You should be familiar with the different methods of collecting population data. Some of the methods used are illustrated below.

Figure 6.1: Methods of collecting population data

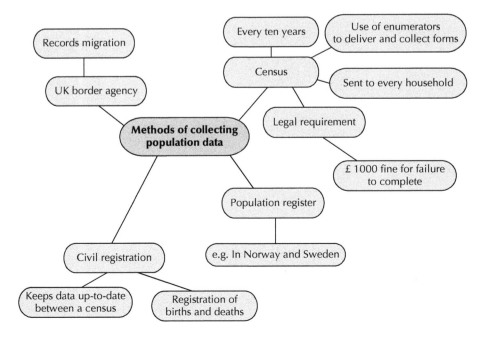

You should be aware of the difficulties involved in collecting population data in certain areas of the world. Figure 6.2 illustrates some of these problems.

Figure 6.2: Problems with collecting population data in a developing country

Explain why collecting accurate population data is more difficult in developing countries than in developed countries.

6 marks

Look at the two answers below. What differences do you notice between them? Think about structure, use of examples and attention to detail.

Answer A

Collecting accurate population data in developing countries can be very difficult due to the sheer size of some countries. The terrain in some areas can also make it difficult to allow accurate data to be collected. Nomads such as people who live in the Sahara Desert move around all the time and have no fixed address. This can make them difficult to find and as a result they may not be counted. **(1 mark)** Tribes in the rainforest have the same problem. Some of these tribes have yet to be discovered and people do not even know they are there.

Another reason for inaccurate data collection could be fear. People may be frightened of the government and may lie on their form. In China, people may lie about how many children they have because of the One-Child Policy. **(1 mark)**

It is also very expensive to carry out a census and some countries are very big, therefore it can be difficult.

2 out of 6 marks

Answer B

Accurate population data is difficult to collect in countries where illiteracy rates are high. Many people may be unable to read the form or will fill it in inaccurately. They may require help from enumerators, which costs money. **(1 mark)** In some countries, people speak many different languages making it expensive to issue a census to everybody. In India for example, there are 22 official languages and over a thousand local languages. **(1 mark)**

Countries with large numbers of migrants, e.g. rural to urban migrants may result in some people being missed or counted twice. **(1 mark)** Similarly if countries have large numbers of people who are homeless, nomadic or live in shanty towns, e.g. Makoko in Lagos, with no official address then they will be difficult for enumerators to track down to deliver a form. **(1 mark)** Mountainous terrain or poor communication links (e.g. the Himalayas) can make it difficult to reach isolated villages, resulting in many people being left uncounted. **(1 mark)**

Carrying out a census may be largely impossible for many countries given the sheer cost involved. Training enumerators, distributing forms and analysing results can be a very expensive process and one that may not be a priority when countries are struggling with health care and education. **(1 mark)**

In addition, many people may lie on their census form in fear of the government, e.g. in China, families may not record the birth of a baby girl as a result of the One-Child Policy. **(1 mark)** Wars, e.g. in Iraq, may also lead to inaccuracies as death rates will be difficult to monitor. **(1 mark)**

6 out of 6 marks

What answer comes out on top?

Answer A is largely descriptive with few developed points. Although size and terrain have been mentioned in the first two sentences, there is no explanation of the difficulties they can cause and so no marks are awarded. However, when discussing 'nomads' and the 'One-Child Policy' the candidate has developed each point and included examples. As a result, 1 mark has been awarded for each factor. The last sentence is again very vague and repeats the point made in the first sentence, without any further development. This is a very basic response and is not of a Higher standard.

In comparison, answer B is very well structured allowing the candidate to see at a glance what they have already included. This helps to avoid repetition and make the best use of time. The candidate has explained each problem in turn, giving examples where necessary. By giving examples, the candidate has added detail to their answer, helping to develop each point. This is an excellent response and one worthy of full marks.

Population structure

Population pyramids show the link between age structure and population change. For the exam, you should be able to:

- Identify what type of country is being represented by a particular pyramid (developed or developing).
- Compare the pyramids of two or more populations.
- Suggest reasons for a given population structure.
- Discuss the social and economic impact of a given population structure for the future population.

Summaries of the population structures for a developed and a developing country are shown in Figures 6.3 and 6.4.

Figure 6.3: Population structure in a developed country

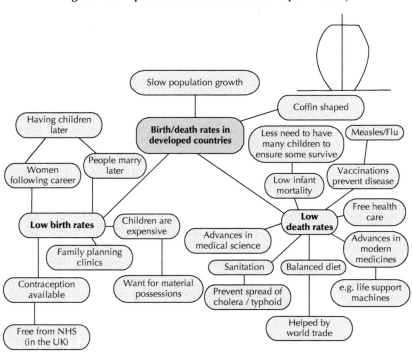

Figure 6.4: Birth and death rates in a developing country

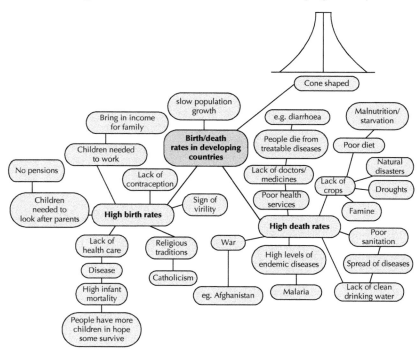

Q Study Figure 6.5. **Discuss** the possible consequences of the 2050 population structure for the future economy of Italy and the welfare of its citizens.

6 marks

Figure 6.5: Projected population pyramid for Italy 2050

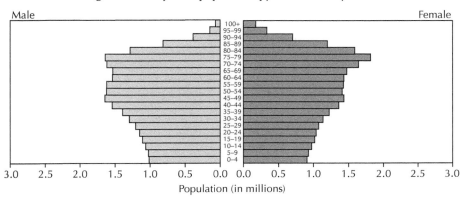

Answer A

With fewer children being born, the demand for maternity care and paediatric nursing will decline, resulting in ward closures and unemployment in these areas. **(1 mark)** *Fewer children will be enrolled in nurseries/schools resulting in forced closure or amalgamation of centres.* **(1 mark)**

The 'greying' of the population may put a strain on current geriatric services, forcing the government to build new health centres and care homes and putting an economic strain on local authorities and central government. **(1 mark)**

However, with a reduction of people in the economically active population there are fewer people to pay for the increasing number of elderly dependents, causing taxes to rise. **(1 mark)** *The age at which workers are able to retire will be forced to rise whilst the government may have to reduce state pensions and encourage private pension plans alongside private health care.* **(1 mark)**

With a reduction in the number of people working, the government may be forced to encourage inward migration to encourage no future shortage in workforce. **(1 mark)** *In times of recession and high unemployment this may lead to racial/religious tension within communities and language barriers could be a problem.* **(1 mark)**

In addition, the government may have to encourage a higher birth rate by increasing maternity/paternity leave and increasing child care benefits. **(1 mark)**

6 out of 6 marks

What makes this a strong answer?

This is a very detailed response, which regularly links back to the question. The candidate has successfully discussed both **economic** and **social** consequences and has explained each point in detail. The structure of this answer is particularly strong, for example:

'With a **reduction in the number of people working** (state problem), *the government may be forced to* **encourage inward migration** (solution to problem)... *this may lead to* **racial/religious tension** *within communities and* **language barriers** (consequence of solution) *could be a problem.* **(1 mark)**'

They have said what the problem is, stated how the government may solve this problem then discussed the possible consequences to the solution. The candidate has used the source well and as a result has produced an excellent answer.

Answer B

With an ageing population, taxes must rise to accommodate the increasing pressures felt by the government. As people are now living longer, people will have to work longer before receiving their state pension and it will be lower than before. **(1 mark)**

A reduction in the number of births will also have negative consequences with regard to the economy. Fewer children will be attending schools and so there will be less need for as many teachers. This will result in many teachers losing their job and they may be forced to retrain in order to find employment. **(1 mark)**

Pressure on large towns to accommodate for the growing number of elderly dependents may result in the creation of shanty towns, to which some residents may object.

2 out of 6 marks

What makes this a weak answer?

This is a very basic response that makes little use of the source provided. Furthermore, the candidate only refers to economic consequences, failing to fully address the question. This is a careless mistake and one that will be detrimental to the overall number of marks awarded. To avoid this, the candidate may have benefited from splitting their answer into sub-sections to help with structuring.

In addition, the last paragraph is irrelevant when talking about a developed country and achieves no marks. Overall, this is a poor answer which does not reflect the standard expected at Higher level.

Migration – people on the move

You should be familiar with both **forced** and **voluntary** migration. You should be able to explain **why** people migrate and the **impact** of migration on society.

Figure 6.6: Push/pull factors of migration

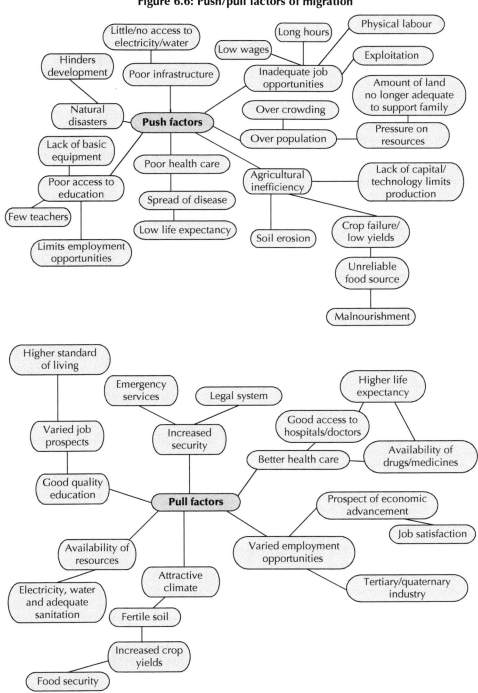

Motives for migrating

When referring to the causes of migration, you should be able to refer to a number of **push** and **pull** factors. **Push** factors work in the migrants' current location whilst **pull** factors are those that attract the migrant to a particular area. These are summarised in Figure 6.6.

Impact of migration

Similarly, the impact of migration can also be subdivided into the impact on the receiving country and the impact on the donor country. The advantages/disadvantages of migration in these areas are summarised below.

Figure 6.7: Impacts of migration

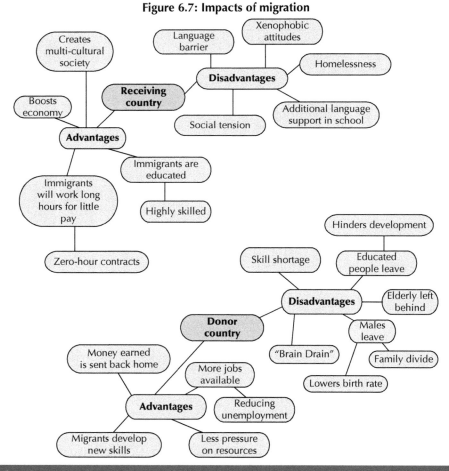

Q With reference to an international migration you have studied, discuss the advantages and disadvantages that the migration has brought to either the donor country **or** the receiving country.

6 marks

Before attempting this question you must be sure of two things:

1. The case study you plan to use.
2. Whether you are going to focus on the **donor country** or the **receiving country.**

In a question where there are multiple parts, **structure** is the key to success. It is a good idea to scribble key words down in the margin to help you as you work through your answer.

For the purpose of this question, you should divide your answer into two distinct paragraphs: **Advantages** and **Disadvantages**. Regardless of your case study, the advantages and disadvantages are likely to be very similar. However, it is the examples that you provide that will set your answer apart.

The answer given below will help you structure your response. Note that the words in **bold** show the factors you should be referring to regardless of the case study. The points that are underlined indicate information specific to the given case study.

First, state your case study, i.e. Polish to Britain. Never assume the examiner knows what you are talking about – state your case study clearly at the beginning of your answer. If you do not refer to a specific case study, marks will be lost.

Advantages (to receiving country – UK)

Many of the immigrants are **educated** *and* **highly-skilled,** *falling into the 20–35-year-old age group. For example about 4000 Polish doctors have been recruited by the NHS. These young professionals are very desirable and contribute well to the* **economy**. **(1 mark)**

> ## Top Tip ✔
>
> In a question that asks you to **discuss**, you must explore both sides of the argument in order to gain full marks, i.e. both advantages **AND** disadvantages must be referred to here if full marks are to be awarded.

> ## Top Tip ✔
>
> It is useful to have sub-headings to help divide your answer. This not only makes it easier for you to ensure each part of the question has been answered, but it also makes it easier for the examiner to mark.

Polish people do not have the same **workers' rights** *as British people and therefore are able to work long hours for little pay.* **(1 mark)** *Many are also employed on* **zero-hour contracts***, which is a big advantage to employers.* **(1 mark)**

Migrants also tend to be very **hard working,** *in fear of losing their job.* **(1 mark)**

The influx of **cheap labour** *has created an extra 0.2% of GDP growth and interest rates 0.5% lower than they would otherwise be.* **(1 mark)**

Migrants help create **multi-cultural societies**, *bringing with them new customs, food, drink and music, e.g. the Polish Club in Glasgow.* **(1 mark)**

Disadvantages (to receiving country – UK)

The **language barrier** *is often a problem, with many Polish people being taken advantage of due to their inability to speak English.* **(1 mark)**

'Poorer' communities resent the incomers for taking entry-level jobs, resulting in **stagnant wages** *for low-paid workers.* **(1 mark)** *As a result* **xenophobic** *attitudes towards EU migrant workers have increased.* **(1 mark)**

Those who are unable to find work are often forced into homeless shelters, e.g. the Cowgate Centre (a 24hr emergency shelter in Edinburgh). It has been recorded that up to 20 Poles are turning up every night, adding to the problem of **homelessness** *in the city.* **(1 mark)**

In addition, the number of children enrolling in Edinburgh schools who require English as an **Additional Language Support** *costs the council more than £1 million a year.* **(1 mark)**

The rates of **British unemployment** *during the recession caused social tension, with many blaming an influx of migrants.* **(1 mark)**

Maximising Marks

For maximum marks to be awarded your answer should contain **detailed explanations**. It is a good idea to include relevant examples to help add detail to your answer, e.g. when answering a question on the advantages/disadvantages of migration, *'with little money, many migrants can have a problem finding accommodation'* on its own would receive no marks, whereas, *'with little money, many migrants can have a problem finding accommodation, e.g. the Cowgate Centre in Edinburgh reports ~20 Polish people turning up every night, adding to the problem of homelessnesss'* would receive 1 mark. Therefore providing examples of a problem/scenario can turn a limited response into a detailed one.

Glossary

Active population: The proportion of a population that is of working age (between 16 and 64).

Birth rate: The number of live births per 1000 people per year.

Death rate: The number of deaths per 1000 people per year.

Dependent population: The proportion of a population not of working age. There are two groups of dependents: *Young dependents* (aged between 0 and 14) and *Elderly dependents* (aged 65+).

Enumerators: People who deliver and collect a census form.

Forced migration: When people are forced to move from their homeland perhaps by civil war, natural hazards, persecution or exploitation.

Infant mortality rate: The number of infants dying under one year old per 1000 live births.

Life expectancy: The average number of years that a person born in a particular country at a particular time is expected to live for.

Migration: The movement of people from one place to another.

Refugee: Someone who has been displaced from their homeland by war, persecution or a natural disaster.

Voluntary migration: When people choose to move from one place to another.

Xenophobic: A fear or strong dislike of people from other countries.

Rural

Topics include:

- The impact of rural land degradation in a rainforest or semi-arid area
- The management strategies employed to control rural land degradation in a rainforest or semi-arid area

Rainforests and semi-arid lands are prone to land degradation. You should be able to **explain** the impact of rural land degradation and **discuss** the management strategies involved in controlling the problem.

Land degradation in the rainforest

Although you will not be asked directly about the human and physical causes of rural land degradation in the rainforest, you should be familiar with these and be able to link them to the management strategies employed.

Figures 7.1 – 7.4 help to summarise rural land degradation in the Amazon Basin.

Figure 7.1: Human causes of rural land degradation in the Amazon Basin

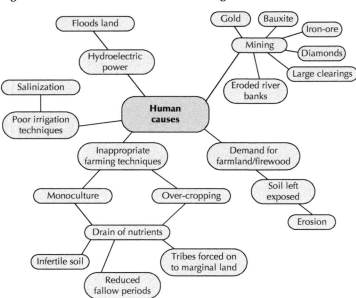

Figure 7.2: Physical causes of rural land degradation in the Amazon Basin

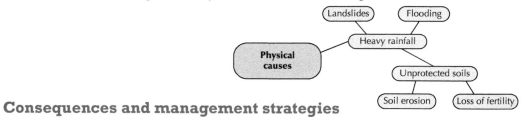

Consequences and management strategies

You should be able to **discuss** the consequences of rural land degradation in the rainforest and **explain** the strategies used to manage the impacts. A summary of these consequences is shown in figure 7.3.

Figure 7.3: Consequences of rural land degradation in the Amazon Basin

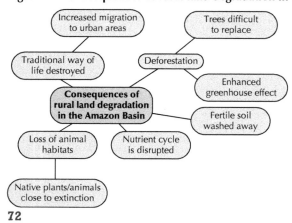

You should also be able to **explain** each of the strategies used to manage land degradation in the rainforest, and comment on their effectiveness. A summary of the strategies used is shown in figure 7.4.

Figure 7.4: Soil conservation strategies in the Amazon Basin

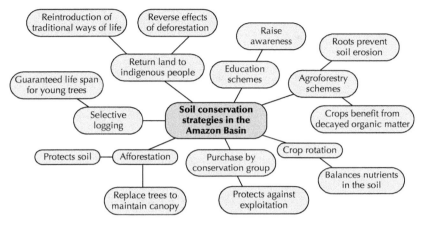

Land degradation in a semi-arid area

Although you will not be asked directly about the human and physical causes of rural land degradation in a semi-arid region, you should be familiar with these and be able to link them to the management strategies employed.

Figures 7.5 and 7.6 summarise the human and physical causes of rural land degradation within a semi-arid region.

Figure 7.5: Human causes of rural land degradation in a semi-arid area

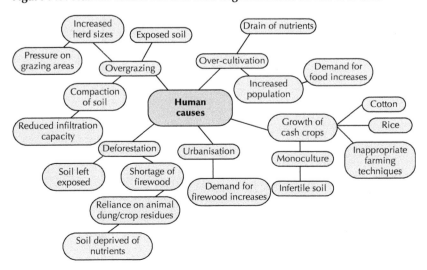

Figure 7.6: Physical causes of rural land degradation in a semi-arid area

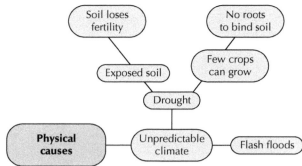

Consequences and management strategies

You should be able to **discuss** the consequences of rural land degradation in a semi-arid area. Some of these are summarised in Figure 7.7.

Figure 7.7: Impact of rural land degradation in a semi-arid area

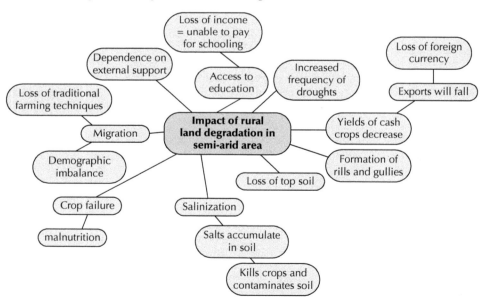

Q For a named semi-arid area you have studied:

(a) Explain the soil conservation strategies used to combat land degradation.

(b) Discuss the effectiveness of each of your chosen strategies.

6 marks

Figure 7.8 illustrates some of the strategies you should refer to.

Figure 7.8: Soil conservation strategies in a semi-arid area

Look at the two answers below. What differences do you notice between them? Think about structure, level of detail and use of examples.

Answer A

Planting new trees helps prevent soil erosion as the roots will bind the soil together and will also act as a windbreak, reducing wind erosion. **(1 mark)** *In places in the Sahel the kad tree has been planted to restore soil fertility and productivity.* **(1 mark)** *This has been effective as its leaves help to shade the soil and provide an additional source of nutrients in the form of decaying leaf litter.* **(1 mark)**

Lines of stone are placed along the contours of the land in an attempt to reduce soil erosion and the loss of water by overland flow. **(1 mark)** *This is very cost effective and communities in Mali and Burkina Faso have seen crop yields increase by as much as 50%.* **(1 mark)**

In order to allow grasses to re-establish themselves they should be given a fallow period of around two years. This means grazing should be controlled and areas should be fenced off on a rotational basis. **(1 mark)** *This prevents long-term damage to the soil. However, fencing can be very expensive and management of grazing areas can be difficult, e.g. in Korr, northern Kenya.* **(1 mark)**

Top Tip

If a question asks you to 'refer to a named area you have studied', marks will be lost for answers that are vague or over-generalised. At Higher level it is not enough to simply name an area. You should include specific examples throughout your answer to demonstrate your case study knowledge.

Grazing can also be controlled through the reduction in the size of a herd. A reduced herd will decrease grazing pressure and the impact of soil compaction. This is very difficult to implement as traditionally the size of one's herd was a sign of masculinity and wealth and therefore herders may be resistant to change. **(1 mark)**

Educating farmers about the causes and consequences of rural land degradation can help change detrimental practices. For example, farmers could be educated on crop rotation or appropriate irrigation techniques such as drip irrigation. **(1 mark)** *However, these alternatives are often more expensive or difficult to manage, resulting in few farmers willing to change their practice.* **(1 mark)**

6 out of 6 marks

Answer B

There are many factors that contribute to rural land degradation. Due to improved health care, population rates are rapidly increasing within the Sahel, putting more pressure on the landscape. As the population increases, there is more demand for food and over-cultivation occurs. This accelerates desertification and decreases the land available to farm. In addition, trees are cut down to produce more available farmland. The removal of trees means the soil is fully exposed to the elements, leading to further erosion.

In some areas, new trees have been planted to help bind the soil together. The leaves will help to protect the soil from rain and they can also act as a wind break, protecting the soil from any further erosion. **(1 mark)** *Areas can also be fenced off to control the movement of animals. If too many animals are trampling over an area, the soil will become compact and will lose fertility. By fencing areas off this will give land time to regenerate whilst still allowing animals to graze the land.* **(1 mark)** *If desertification takes hold and land becomes too infertile to farm, communities can become too dependent on aid from other countries.*

2 out of 6 marks

What answer comes out on top?

Choosing to combine parts (a) and (b), answer A explains each strategy in detail before discussing its effectiveness. Structuring an answer like this allows you to see what you have already included, avoiding repetition, and helps to ensure you don't miss out vital points. Providing examples is an excellent way of showing the examiner you know your case study and may help you pick up an extra mark.

However, having achieved 2 out of 6, answer B is a classic example of a candidate who has 'dived' into the question. The first paragraph looks at the causes of rural land degradation and therefore has no relevance to the question. The candidate has then used the second paragraph to explain some of the strategies used, yet has made no reference to the effectiveness of these strategies. Although the strategies mentioned were relevant and earned the candidate two marks, this answer is very limited with regard to the explanation of strategies employed. In addition, the last sentence refers to a consequence of land degradation. Again, this is not relevant to the question asked and therefore no additional marks have been awarded.

Maximising Marks

At Higher level, it is not enough to simply give descriptive points. Answers should be detailed and give examples where necessary. For example, when discussing soil conservation strategies in a rainforest area, for one mark you should give one detailed explanation, e.g.

'Selective logging ensures trees are only felled when they reach a particular height, allowing the forest to regain full maturity after around 40 years while giving young trees a guaranteed life span.' **(1 mark)**

OR provide a limited description/explanation of two factors, e.g.

'Agroforestry schemes allow crops and trees to be grown on the same area of land to protect and conserve the soil, while education schemes are set up to teach people about the consequences of their actions.' **(1 mark)**

Glossary

Afforestation: The planting of trees.

Agro-forestry: The growing of trees and crops on the same land.

Degradation: The deterioration of an area over time.

Desertification: The spread of desert-like conditions into neighbouring regions.

Exploitation: The use of a resource for profit.

Extinction: When there are no surviving individuals of a particular species.

Indigenous people: People who are native to an area.

Selective logging: When trees are only felled after reaching a particular height.

Soil erosion: The washing or blowing away of topsoil leaving behind infertile land.

Section 8: Urban

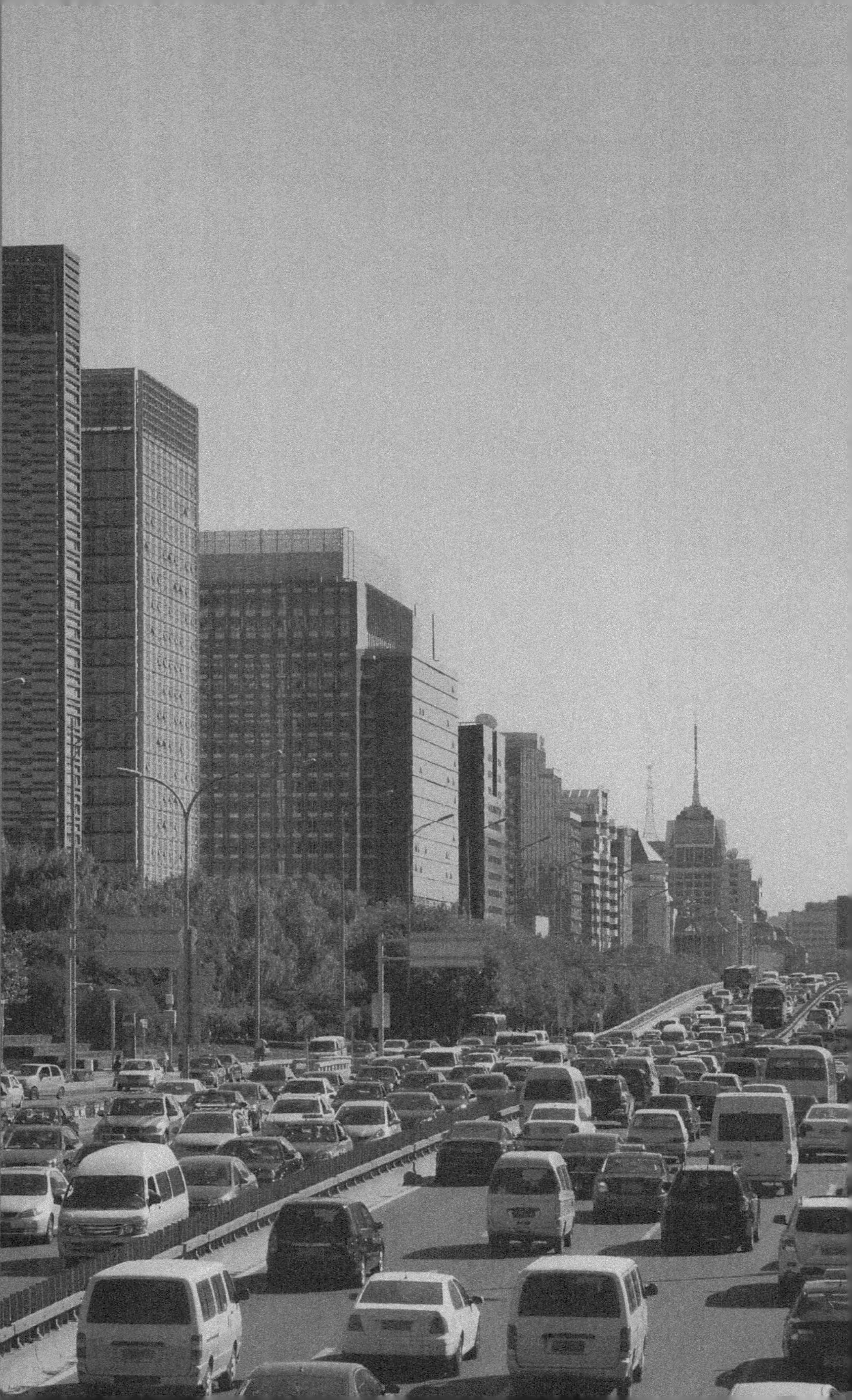

Urban

8

Topics include:

- Need for management of recent urban change (housing and transport)
- Management strategies employed
- Impact of management strategies

In developed and developing cities, problems and changes frequently occur in urban areas. You should be able to explain these problems and how they may be resolved.

Management of urban change

Traffic congestion

Traffic congestion is a major issue in cities around the world. You should be familiar with the causes of traffic congestion and be able to explain the solutions to help solve the problem. A Summary of the solutions to traffic congestion is shown in figure 8.1 below.

Figure 8.1: Solutions to traffic congestion (with reference to Glasgow as an example)

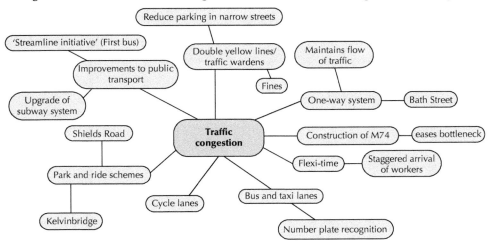

> **Q** For Glasgow or any other named city in a **developed** country, **explain** the measures used to reduce problems of traffic management within the Central Business District.
>
> **5 marks**

Answer A

In Glasgow, the Clyde Tunnel and Kingston Bridge were built to ease traffic congestion across the River Clyde and direct traffic away from the city centre. A further attempt was made to ease this bottleneck by way of the M74, which was completed in 2011. **(1 mark)**

Improvements have been made to public transport systems to encourage more people to leave their car at home. This includes the modernisation of the subway system and redevelopment of Buchanan Bus Station. **(1 mark)** First buses have also introduced a new service known as Streamline. This is an enhanced bus service that aims to provide a more reliable, accessible and affordable journey around Glasgow. **(1 mark)** Bus and taxi lanes have also been created to allow public transport to move more quickly, creating a more reliable service. Number plate recognition cameras are in place to fine those who incorrectly use these lanes. **(1 mark)**

Park and ride schemes have been developed, where people park their car at the edge of the city and take a train or bus into the city centre, e.g. Kelvinbridge subway station offers extensive parking and discounts to passengers. This helps to reduce the number of cars entering the city. **(1 mark)**

One-way systems help to keep the traffic moving (e.g. Bath Street) and double yellow lines/traffic wardens help to reduce on-street parking, creating more space on narrow city roads. **(1 mark)**

Flexi-time is another approach that has been adopted by many businesses around the city to encourage a staggered arrival of workers. Flexi-time allows workers to choose when they start work in the morning, helping to reduce the flow of traffic during rush hour. **(1 mark)**

5 out of 5 marks

What makes this a strong answer?

This candidate has referred to a wide range of measures, giving this answer depth while avoiding repetition. The inclusion of named examples demonstrates the candidate's case study knowledge and adds detail to their answer. Overall this response is well structured and full marks are well deserved.

Answer B

Traffic congestion is caused by too many people trying to access the same place at the same time. This is particularly common at rush hour when lots of people are trying to get into the *city for work. To solve this, the use of public transport has been encouraged. One-way streets have also been put in place to keep the flow of traffic moving smoothly.* **(1 mark)** *Park and ride schemes, flexi-time and increased parking charges have been introduced to ease the pressure on the roads. These schemes have been successful and have reduced the number of cars on the road.*

1 out of 5 marks

What makes this a weak answer?

The first two sentences here refer to causes of traffic congestion, failing to answer the question. The candidate makes a good point with *'public transport has been encouraged'* yet includes no further explanation. Opportunities have also been lost as the candidate then lists some of the measures employed, offering no additional detail. Statements like *'these schemes have been successful and have reduced the number of cars on the road'* are simplistic and become meaningless when left undeveloped. The lack of explanation or development of points has proved costly for this candidate, resulting in 1 mark out of 5.

Urban sprawl

Urban sprawl refers to the constant expansion of towns and cities into the countryside. You should be familiar with the problems caused by urban sprawl and be able to discuss how these problems may be resolved. The problems and solutions to urban sprawl are summarised in figures 8.2 and 8.3.

Figure 8.2: Problems caused by urban sprawl

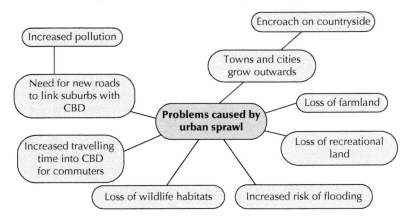

Figure 8.3: Solutions to urban sprawl

Figure 8.3: Solutions to urban sprawl

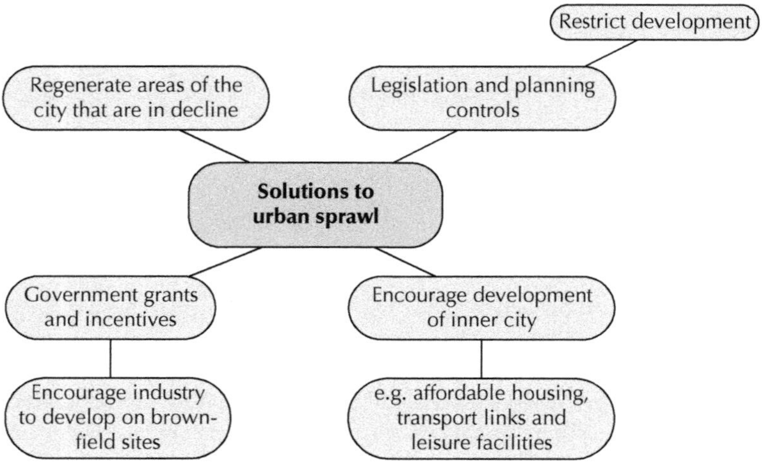

Q Referring to Paris or any other city you have studied in a developed country:

(a) explain the problems caused by urban sprawl

(b) discuss how these problems may be solved.

6 marks

Look at the two answers below. What differences do you notice between them? Think about the level of detail, relevance of information and structure of answer.

Answer A

(a) *Urban sprawl occurs on the rural/urban fringe, often resulting in towns and cities joining up. New housing and industrial estates are often the cause of this where residents and businesses are looking for large areas of cheap, unpolluted land. Towns such as Cumbernauld were built at the edge of the city to relieve the pressures of over-crowding within inner city areas. Today, new housing developments continue to add to the problem, further expanding the city boundary. Developments like this destroy animal habitats and could lead to their extinction.*

(b) *Glasgow City Council have tried to reduce the effects through the regeneration of inner city areas. They have encouraged people to build on sites that have been left derelict to limit the amount of 'new' land that is being destroyed.* **(1 mark)**

1 out of 6 marks

Answer B

(a) *As towns and cities develop, more and more fertile farmland is lost, reducing the area of land available to feed a growing population.* **(1 mark)** *Recreational land is also lost and flooding issues may increase if developments are built on floodplains.* **(1 mark)**

As green spaces are reduced, wildlife habitats will increasingly come under threat, which can have an additional impact on the quality of crops produced. **(1 mark)**

Suburban commuters will be faced with longer travelling times into the CBD and traffic congestion will be increased as bottlenecks will occur, e.g. over the Kingston Bridge. **(1 mark)** *New roads will have to be built, e.g. the M74, which will increase levels of air pollution, further contributing to greenhouse gas emissions.* **(1 mark)**

(b) *Some of these problems can be solved through legislation and planning controls put in place by the local government, which will restrict development in green belt areas.* **(1 mark)** *Redevelopment of the inner city should be encouraged before digging up the countryside. For example, in Glasgow, inner city areas like the Gorbals have been developed with modern, affordable housing, good transport links and leisure facilities to try and encourage residents to stay in the city.* **(1 mark)** *In addition, government grants and incentives have been used to encourage industry to develop on brownfield sites. This not only protects the countryside but regenerates areas of the city that were in decline.* **(1 mark)**

6 out of 6 marks

What answer comes out on top?

In answer A, the candidate has dived into the question and as a result, has misunderstood what it is asking them to do. For part (a), the candidate has looked at general causes of urban sprawl before mentioning the problems very briefly. This has not answered the question and no marks have been awarded. Part (b) makes an attempt at discussing how these problems can be solved. However, this is very brief and fails to link with part (a). One mark has finally been awarded for a limited description of two points. Yet, on the whole, detail is limited, producing an answer of little value.

Answer B on the other hand regularly links back to the question and fully explains the problems caused by urban sprawl. The candidate has looked at a wide range of problems and has included examples where necessary. Although both parts have been answered separately, the candidate has still linked parts (a) and (b) together, resulting in an answer that is clear, logical and precise.

Urban problems in developing countries

Due to rapid population growth, shanty towns are a major problem within developing countries. You should be familiar with the problems within a shanty town and be able to explain the methods used to tackle these problems while commenting on their effectiveness. Figure 8.4 summarises some of the main problems.

Figure 8.4: Problems in shanty towns

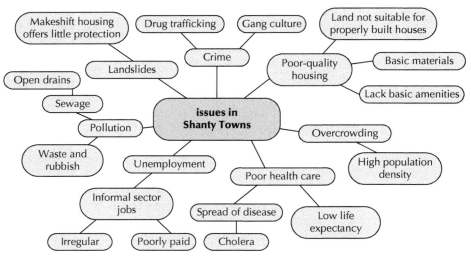

For Rio de Janeiro or any other named city in a **developing country, explain** the methods used to tackle problems in shanty towns and comment on their effectiveness.

5 marks

Answer A

The living conditions in shanty towns are very poor. Often they do not have running water or proper sanitation schemes. People are often forced to turn to crime, which makes these areas very dangerous for people. Sometimes, locals will run tours around the shanty towns for tourists. This can be dangerous and tourists have been known to get mugged. Governments don't like shanty towns but there is often no other solution as if you demolish them the people will have to move elsewhere. This would be an example of hiding the problem as oppose to solving it.

Some authorities have improved shanty towns by installing pipes and sanitation. Some shanty towns now even have proper roads that link with the cities. People can use these roads to travel to the cities for work and they can then use the money earned to improve their standard of living, reducing the need for residents to resort to crime. **(1 mark)** Sometimes, the shanty towns that are near high-end business parks or within tourist areas will be flattened and the residents are moved to the outskirts of the city. This can be a good idea sometimes but some people might disagree.

1 out of 5 marks

What makes this a weak answer?

This candidate has been careless with the question resulting in an answer that is vague and over-generalised. The first paragraph solely describes the general problems within a shanty town and makes no reference to how to solve them.

The second paragraph is slightly better, with the candidate referring to some methods used to tackle the problems. However, the information given is largely descriptive with only one mark being awarded for a combination of limited descriptive and explanatory points. Statements like *'this can be a good idea sometimes but some people might disagree'* are particularly weak and add nothing to the answer on a whole. Overall this is an answer that contains very little detail, achieving 1 out of 5.

Answer B

In places like Rio de Janeiro, self-help schemes have been introduced to encourage people to develop their skills and improve their standard of living. **(1 mark)** *Local authorities provide materials such as bricks, cement and glass and encourage local residents to work together, transforming the favelas into small towns/cities, e.g. Rochina.* **(1 mark)** *In Rochina, self-help schemes have transformed the slums into low-quality housing with basic facilities like electricity and sewage systems.* **(1 mark)** *However, given the relief of the land, the vast number of people and the amount of funds available, the success of these schemes is limited.* **(1 mark)**

Local Authority Programmes have been set up to relocate residents, e.g. the Favela Bairro Project. Brick houses have been built with sanitation pipes and running water and services include refuse collection, street lighting and emergency services. **(1 mark)** *In some areas there are shops, cafes and entertainment facilities run by local people. This provides jobs, helping to decrease levels of crime.* **(1 mark)** *However, out of the 600 favelas in Rio de Janeiro, only 60 have benefited from this program as there are not enough funds available.* **(1 mark)**

High-rise tower blocks have been built in the suburbs to provide high-density housing away from the tourist spots of the city. This simply relocates the problem and can cause tension between different social groups in the city. **(1 mark)**

Charities and aid-workers work with people in shanty towns to help improve their standard of living. The Developing Minds Foundation works in Rio de Janeiro to build schools. **(1 mark)** *This helps to improve the literacy rates of children, enhancing their life choices as they grow up.* **(1 mark)**

5 out of 5 marks

What makes this a strong answer?

This is a very detailed response that clearly explains the methods used to tackle problems within shanty towns. The candidate has used a range of specific examples rather than brief references, which adds depth to the answer. Each method has been evaluated and the candidate has successfully commented on how effective each method has been. This response is typical of one from an A candidate and deservedly achieves 5 out of 5.

Maximising Marks

In order to gain full marks, you must do exactly what the question is asking you to do. For example, if a question asks you to *explain methods used* **then** *comment on their effectiveness*, you must refer to both parts of the question for full marks to be awarded, e.g.

'High rise tower blocks have been built in the suburbs to provide high-density housing away from the tourist spots of the city (explanation of method). *This simply relocates the problem and can cause tension between different social groups in the city* (comment on effectiveness).

Top Tip

When commenting on the effectiveness of a method/ strategy you must use knowledge from your case study to back it up. Simply stating that something has been successful/ unsuccessful is meaningless unless accompanied by supporting statements.

Glossary

Conurbation: Formed when cities and towns spread outwards and join up with each other, creating one continuous urban area.

Decentralisation: When businesses move out of the CBD to a less congested location out of town.

Derelict building: A deserted or abandoned building.

Green belt: An area of land surrounding a city/town that is designed to protect rural areas from urban sprawl.

Rural/urban fringe: The boundary between a city or town and the countryside.

Self-help schemes: A project, often in developing countries, where local people are given tools and training to learn new skills in order to help themselves and their communities.

Urban sprawl: When towns and cities grow outwards and encroach on the countryside.

River Basin Management

Topics include:

- Physical characteristics of a river basin (climate, rock type, vegetation, river flow)
- Need for water management
- Selection and development of sites
- Consequences of water control projects

Need for water management

Water management is needed in certain areas due to a number of human and physical factors such as rainfall patterns, average temperatures and population growth. You should be able to look at a number of sources and explain why a water management project is needed in a given area. Some of these factors are summarised below.

Figure 9.1: Need for water management

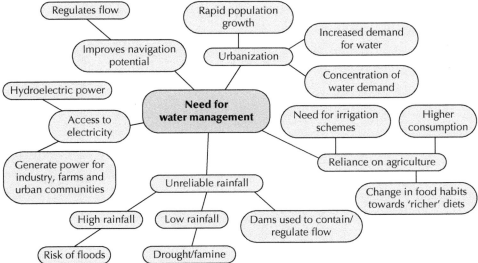

Q With reference to figures 9.2 and 9.3 **explain** why there is a need for water management in Egypt.

5 marks

Figure 9.2: The Nile Basin

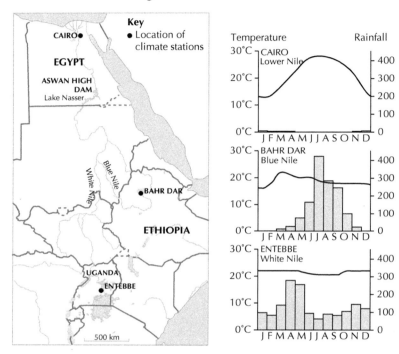

Key
● Location of climate stations

CAIRO
EGYPT
ASWAN HIGH DAM
Lake Nasser
White Nile
Blue Nile
● BAHR DAR
ETHIOPIA
UGANDA
● ENTEBBE
500 km

Temperature / Rainfall

CAIRO Lower Nile

BAHR DAR Blue Nile

ENTEBBE White Nile

Figure 9.3: Population of Egypt (1950–2050)

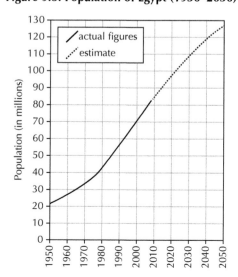

Population (in millions)

— actual figures
···· estimate

Answer A

In Cairo there is no rainfall between May and September. However, in Bahr Dar there is no rainfall between December and February, then it rains heavily for the other months of the year. Entebbe sees rainfall all year round with the heaviest falling in April. As there is a lot of rain, some form of water management scheme needs to be put in place. If people are growing crops or cooking they will need a water supply. At the moment in Egypt, the people are not getting a steady water supply, resulting in deaths.

0 out of 5 marks

> ### Top Tip ✔
>
> In order to answer this question fully you must **explain** the need for water management within the Nile river basin using the relevant prompts from the resources provided.

What makes this a weak answer?

The first three sentences here are purely descriptive with no marks being awarded. The candidate has made poor use of the resources, referring only to the climate graphs. This has proved detrimental to their overall mark. Although the candidate has described the rainfall on the graphs they have not linked the information to the question. By stating that some form of water management scheme is needed the candidate has simply reworded the question, making no attempt to develop their point or give any form of explanation. This answer is vague and achieves no marks.

Answer B

There is very low rainfall in Egypt all year round meaning that water management is needed to ensure that people have water for activities such as washing and drinking. **(1 mark)** *High rainfall further south within the Nile Basin between April and September shows that flooding is a threat and so dam building is required as a flood control measure.* **(1 mark)**

The rapid population growth predicted for Egypt in the future suggests that demand for water from the growing population will be high and will continue to increase. This shows a need to regulate water flow and storage. **(1 mark)**

As the population increases, the demand for food increases. This means more water is needed for irrigation purposes in order to feed the population. **(1 mark)** *In addition, as the number of people increases the need for hydroelectric power becomes more important as energy is needed to power expanding cities and industry.* **(1 mark)**

5 out of 5 marks

What makes this a strong answer?

This candidate has used the resources well to explain the need for water management in Egypt. Each source has been analysed before applying the information to the question. The inclusion of this information adds depth to the answer, providing detailed explanations and developed points. This is a very strong, well-structured answer and is worthy of full marks.

Maximising Marks

If you are asked to explain the need for a water management project, you should use all resources provided within the question. No marks will be awarded for solely descriptive points, e.g.

'In Cairo there is no rainfall between May and September'.

For 1 mark to be awarded, you need to **apply** the information from the sources to the question, e.g.

'High rainfall further south within the Nile Basin between April and September (information from sources) *shows that flooding is a threat and so dam building is required as a flood control measure* (information applied to question)' **(1 mark)**

Selection and development of sites

There are many physical and human factors which must be taken into account when selecting and developing a site for a dam. These factors are summarised below.

Figure 9.4: Selection and development of sites

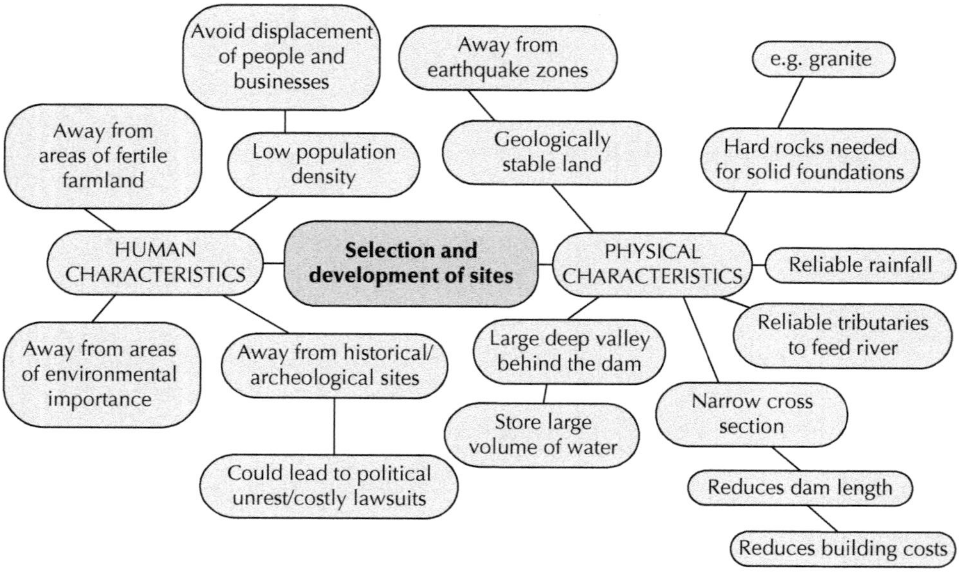

Consequences of water control projects

For a named water management project, you should be familiar with the social, economic and environmental benefits as well as the adverse consequences. With reference to the Colorado river basin, these consequences are summarised below.

Figure 9.5: Consequences of water control projects

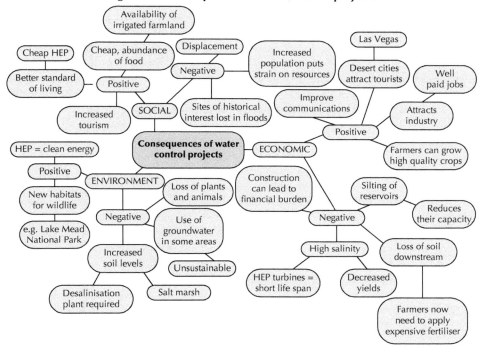

Q **Evaluate** the social, economic and environmental **benefits** of any major water control project you have studied.

5 marks

Hint: for this question, answers must discuss positive consequences of the water control project. No marks will be awarded for negative consequences.

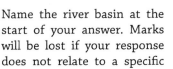

Top Tip

Name the river basin at the start of your answer. Marks will be lost if your response does not relate to a specific water control project.

Answer A

Environmental benefits

Through the water management project on the Colorado river, new wildlife habitats have been created. Species such as the egret are now able to thrive in lakes such as those in Lake Mead National Park. **(1 mark)** New flood control measures now mean people living downstream are no longer in danger of constant floods. In addition, the use of hydroelectric power is a clean source of energy and therefore does not contribute to the greenhouse effect. **(1 mark)**

Social benefits

People in desert cities such as Las Vegas can enjoy a better standard of living as a result of cheap hydroelectric power projects that enable the use of air conditioning. **(1 mark)** People are now able to have swimming pools and landscaped gardens due to the availability of water, and more than 9 million visitors per year are attracted to the area for water-sports and fishing, e.g. around Lake Mead. This results in improved services for locals. **(1 mark)**

Economic benefits

As a result of cheap hydroelectric power, industry has been attracted to areas such as Arizona, creating well-paid employment opportunities and boosting the local economy. **(1 mark)** The availability of water for irrigating farmland has enabled agribusiness-style farming to develop, covering over 2 million acres in California and Arizona. **(1 mark)** Tourism is now a massive industry in the area, with the Hoover Dam attracting 700,000 people every year and with improved communications, aluminium mining in California is also able to take full advantage of the cheap power available. **(1 mark)**

5 out of 5 marks

What makes this a strong answer?

By splitting this answer into distinct sections, the candidate has given a very detailed, well-structured response that has fully addressed all aspects of the question. Splitting the answer up like this not only makes it easier to write, but also makes it easier for the examiner to mark. Through the use of named examples, the candidate has demonstrated their case study knowledge, helping to develop their points and bring relevance to their answer. No negative points have been made, which shows the candidate has read the question carefully, allowing them to make the most of the time available.

Answer B

Industry has been attracted to areas within the Colorado basin since the new water management schemes were installed. Industry is able to thrive in areas such as Phoenix due to the availability of cheap hydroelectric power. This provides jobs for local people. **(1 mark)** Tourists are attracted to the area to see the Hoover Dam. This brings in a lot of money to the area and helps boost *the economy.* **(1 mark)** *However, the cost of construction and maintenance often counteracts these benefits, leaving a financial burden on those who funded the projects.*

In areas that receive little rainfall and are prone to drought, the water management project is a great advantage. In places like Phoenix people are now able to have nice gardens. However, the ever-expanding population continues to put pressure on these resources, reducing their positive impact.

2 out of 5 marks

What makes this a weak answer?

This answer starts off on track, with the candidate achieving two marks in the first paragraph. Each mark has been awarded for a limited explanation of two factors. However, the candidate then moves on to discuss the negative impacts of the scheme, which achieves no marks. Although this candidate has included named examples within their case study area, the use of *Phoenix* has been repeated. Repeating an example adds no value to an answer and is a poor use of time.

Furthermore, this answer only refers to social and economic effects with no reference to the environmental benefits. This reflects poor exam technique and a better structure may have prevented this. Overall this response has not fully answered the question and achieves 2 out of 5.

Top Tip

One mark will be awarded if you refer to two specific named examples within your case study area, e.g. a named city or species.

Glossary

Aqueduct: a man-made channel that carries water from one place to another.

Dam: A barrier built across a river to hold water.

Hydroelectric power (HEP): Renewable energy source generated by running water.

Reservoir: A lake used for the storage and regulation of water.

River basin: The land area drained by a river and its tributaries.

Tributary: A small river/stream that runs into a larger river.

Section 10: Development and Health

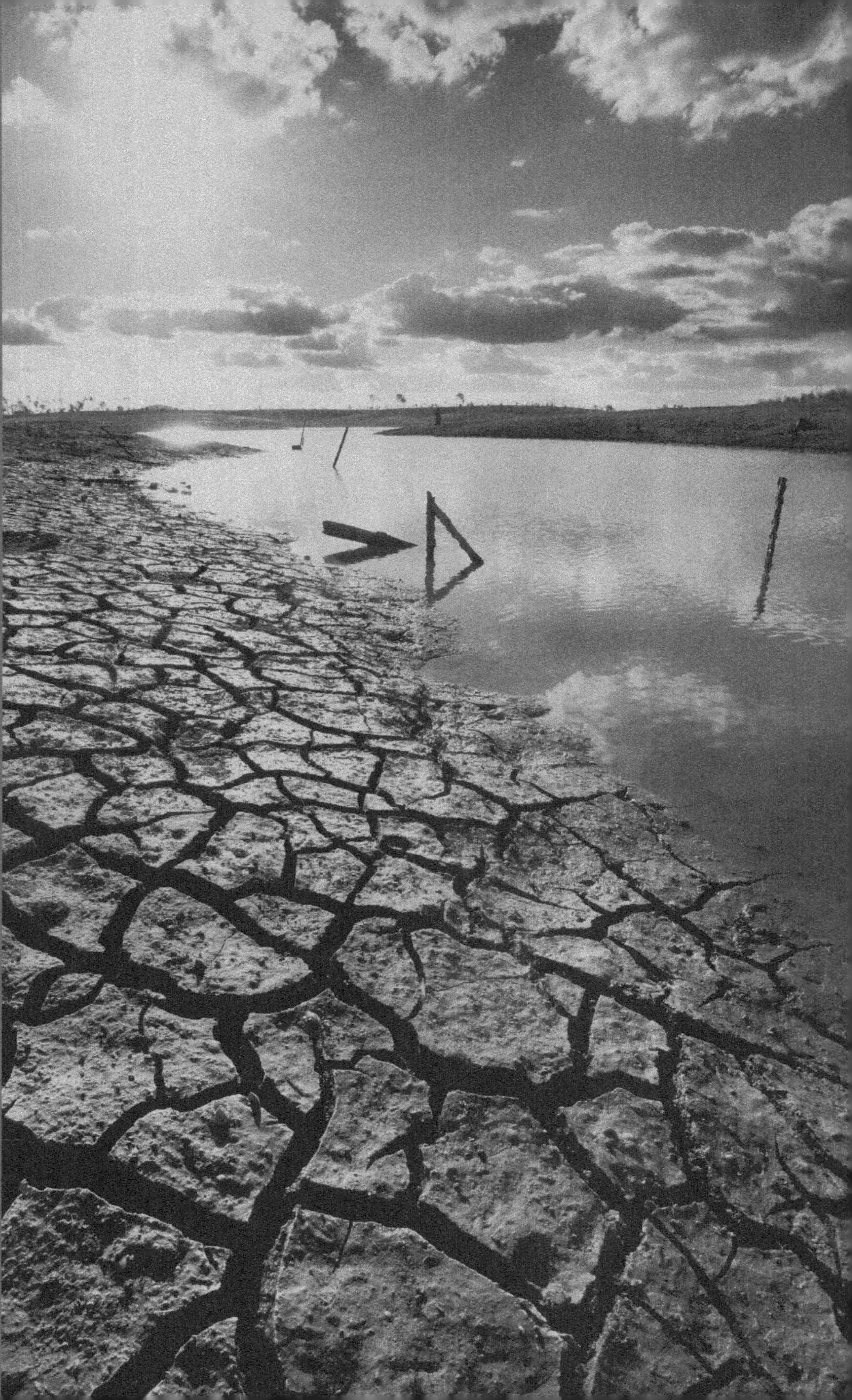

Development and Health

Topics include:

- Validity of development indicators
- Differences in levels of development **between** developing countries
- Study of a water-related disease: causes, impact, management
- Primary Health Care strategies

Development indicators

Development refers to any improvement in the standard of living of people in a specific country. It is measured using **development indicators**. You should be able to explain:

- Social and economic indicators
- Problems of using these indicators
- Composite indicators

Figure 10.1 summarises development indicators.

Figure 10.1: Development indicators

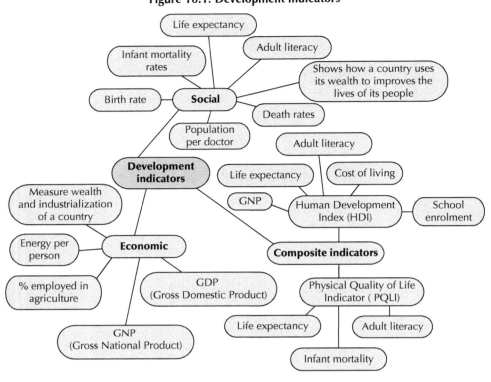

Differences in levels of development between developing countries

Q **Account for** the wide variations in development that exists **between** developing countries. You should refer to **named countries** you have studied. **6 marks**

Top Tip ✔

Before attempting a question like this, take time to scribble down the main factors that affect levels of development as shown below. This will keep you focused and ensure you don't miss out key points.

Top Tip ✔

Remember to link each factor with a named country, e.g. mountainous areas such as Nepal. This will add depth to your answer and could help you pick up marks.

Figure 10.2: Factors which affect development

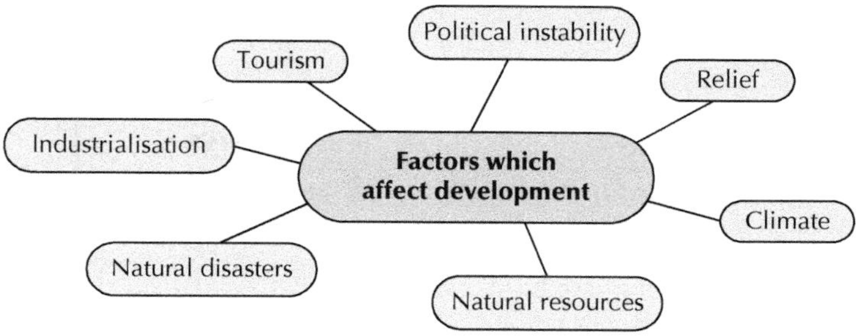

Answer A

Poor climate can hinder development, as areas that are very dry make farming difficult, leading to famine. This can lead to desertification, e.g. in Arid Sahel. **(1 mark)** *Areas like Chad and Mali suffer from drought, therefore it is difficult to grow crops. Northern parts of Brazil (Amazonia) suffer from an inhospitable rainforest climate where heavy rainfall causes poor quality farmland and makes it difficult to build.* **(1 mark)**

Mountainous areas, e.g. Nepal, have a lack of flat land for growing crops or for allowing industry to develop. Finding land suitable for building transport links can also be an issue hindering development. **(1 mark)**

Natural resources bring great wealth to countries such as Saudi Arabia and United Arab Emirates with their oil and gas reserves. This allows them to trade with other countries and bring in money. **(1 mark)** *However, money is not always spent on the people and is controlled by a small percentage of the population.* **(1 mark)**

Natural disasters such as earthquakes, floods and drought can hinder development. Areas prone to drought are associated with famine, such as Sudan and Ethiopia. **(1 mark)** *Floods can destroy farmland, e.g. in Bangladesh, and earthquakes can destroy infrastructure leading to unemployment, e.g. in Haiti.* **(1 mark)**

Industrialisation of countries such as those on the Pacific Rim, e.g. South Korea, Taiwan and China has allowed for the establishment of wide industrial base, e.g. steel and electrical goods, through their entrepreneurial skills and cheap workforce. **(1 mark)** *The money earned from industry and exports can be spent on education and health care, improving literacy rates, infant mortality and life expectancy.* **(1 mark)**

Tourism has helped with development in countries such as Thailand and Sri Lanka. The tourism industry has brought in foreign currency and provided job opportunities. **(1 mark)**

Internal problems can slow down development. Political instability and civil wars in Sudan, for example, have meant that much of their spending is focused on military budgets and arms, instead of education and health care. **(1 mark)** *It has also created large groups of refugees who become dependent on aid, e.g. Rwanda and Zimbabwe. Foreign countries will not invest in these areas so the country fails to develop and this leads to low GDP.* **(1 mark)**

6 out of 6 marks

What makes this a strong answer?

This is a very clear answer with a solid structure. The candidate has taken each factor in turn before explaining how it directly affects the development of a country. By referring to a wide range of factors, the candidate avoids repetition and the vast use of examples adds detail to the answer. This is an excellent response, and one which contains more than enough to achieve full marks.

Answer B

Countries with good climates will attract a lot of tourists. This will bring in a lot of money to the area and will provide jobs for locals. This will help local businesses to develop and will allow local economies to grow. **(1 mark)**

Countries like Brazil offer lots of different opportunities because they are so big. The south-east of Brazil is so flat, encouraging industry to build there. The road network is also good here which allows people and goods to move around easily. Fertile soils in this region allow crops like coffee to grow, which brings in lots of money to the area. **(1 mark)**

Oil can make countries rich. Oil is in demand all around the world, helping countries to develop. However, sometimes resources like oil can be detrimental to a country's development.

Natural disasters can prevent a country from developing. If landscapes are regularly flattened, countries can find it difficult to regulate trade and sell goods. **(1 mark)**

The relief of a country also has a role to play. Soils are thin in mountainous areas, making it difficult to grow things.

3 out of 6 marks

What makes this a weak answer?

This is a fairly generalised answer, with the candidate giving some reasons for variations in development. In paragraph two, reference is made to Brazil where the candidate demonstrates good knowledge of the area. Had the candidate included similar examples throughout the answer, more marks would have been awarded.

In paragraph three, although the candidate has correctly identified natural resources as a reason for variations in development, they follow this up with vague statements like *'oil is in demand all around the world, helping countries to develop'*. With no reference to a specific country and no explanation for how it contributes to a country's development, this point is over-generalised and does not address the question. Similar points have been made in the last paragraph where opportunities have been lost due to a lack of explanation.

Given the candidate has only referred to one named country and development is limited, no more than half marks can be awarded.

Water-related disease

For any water-related disease, you should be familiar with its causes, impact and management strategies.

Malaria

Q For any water-related disease you have studied:

(a) Explain the measures that can be taken to combat the disease.
(b) Evaluate the effectiveness of these measures.

10 marks

You should be familiar with the physical and human factors that contribute to the spread of malaria. These are summarised below.

Figure 10.3: Physical and human factors of malaria

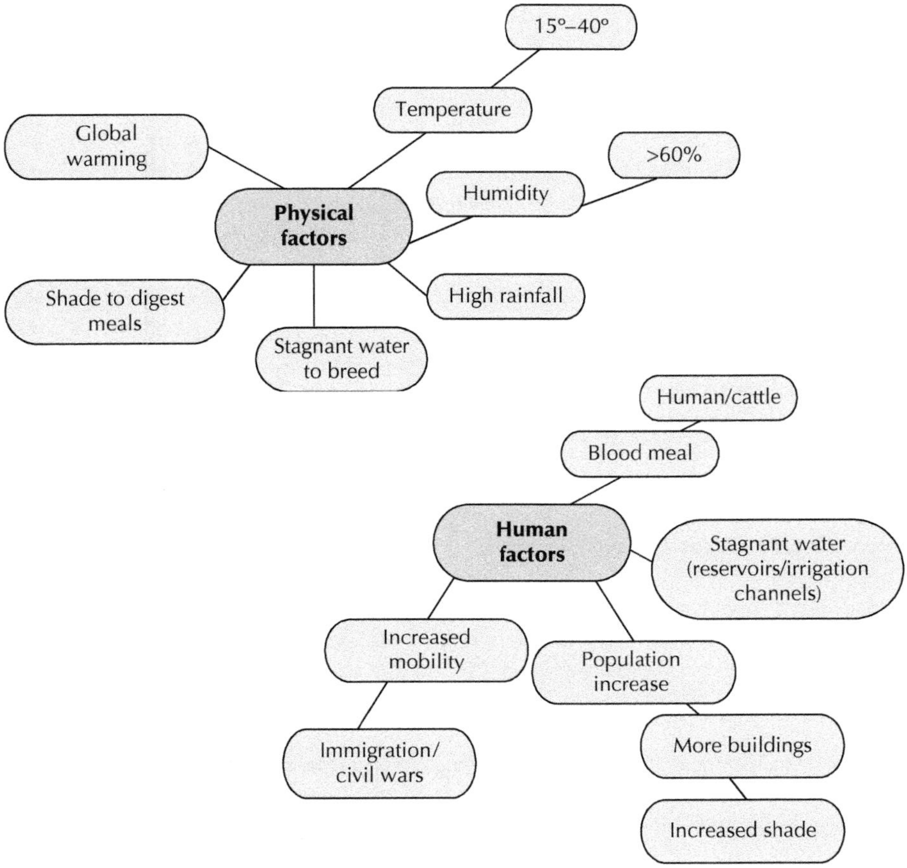

Maximising Marks

When a question is combined like this, you may choose to answer parts (a) and (b) separately or together. However, a maximum of 6 marks will be awarded for any one section.

In order to gain marks for part (a), you must **explain** how each method actually helps to control the disease and not simply describe or list different methods.

Similarly for part (b) **evaluation** points should be developed, e.g. it is expensive to research, therefore making it too expensive for some developing countries.

Answer A

Draining areas of stagnant water such as puddles and swamps every 7 days can reduce the number of breeding grounds for the mosquitoes to lay their eggs. **(1 mark)** *However, in tropical areas there can be heavy rainfall every day so new areas of stagnant water can appear all the time. This method can be very time consuming and is impossible to be effective on a large scale.* **(1 mark)**

Spraying insecticides such as DDT on breeding grounds and in people's homes killed the mosquitoes and their larvae as it destroyed their nervous system. **(1 mark)** *This was easy to do but had to be done thoroughly to be effective and over time mosquitoes have developed resistance to the insecticide. DDT was also harmful to the environment and has since been banned, but now newer insecticides such as malathion have been developed.* **(1 mark)** *As this is petrol-based it is less harmful to the environment but homes and breeding grounds have to be sprayed more frequently and this makes it expensive to use. It also stains the walls of people's houses a horrible yellow colour and villagers complain about the smell, so some don't want it sprayed in their homes.* **(1 mark)**

A newer method is to impregnate coconuts with Bti bacteria. After a few days the fermented coconuts can be broken open and thrown into mosquito-infested ponds. The larvae eat the coconuts and the bacteria destroy their stomach linings. **(1 mark)** *This is a cheap and environmentally friendly way to reduce the number of mosquitoes, as the bacteria will not harm humans or the environment. Coconuts grow well in tropical areas and two or three coconuts will 'control' a small pond for 45 days.* **(1 mark)**

One successful way of controlling the spread of malaria near human settlements in Southeast Asia is to add fish, e.g. the Muddy Loach, to areas of stagnant water. The fish eat the larvae so people are less at risk of catching malaria. **(1 mark)** *The fish are easy to breed and are also a vital source of protein for the people.* **(1 mark)**

Drugs such as Chloroquine and Malarone can be given to people to stop then catching malaria or to treat malaria sufferers. Although it is cheap, mosquitoes have become immune to Chloroquine over the years. Malarone is a newer drug and is around 98% effective. **(1 mark)** *Another newer* drug, Quinghaosu, first discovered in China from the Artemisia plant, is now being developed and initial trials have shown that it is effective in helping malaria sufferers. **(1 mark)** *The drug reacts with the iron in the parasite and kills it before it has time to adapt. However, anti-malaria drugs often have bad side effects such as nausea and headaches.* **(1 mark)**

Education is an effective, simple and cheap way of tackling malaria when done through Primary Health Care programmes. People can be given advice that helps prevent them from catching malaria. An effective yet simple way is to sleep under an insecticide-treated bed net. Mosquitoes are most active between dusk and dawn and nets protect children and adults while they sleep. **(1 mark)**

10 out of 10 marks

Note: '**(1 mark)**' indicates where marks have been awarded for explaining the method [part (a) of the question] while '**(1 mark)**' indicates a mark for evaluating the effectiveness [part (b) of the question].

What makes this a strong answer?

This candidate has chosen to combine parts (a) and (b), resulting in a clear, detailed response. Each paragraph clearly explains the method used **then** evaluates its effectiveness, regularly linking back to the question. Throughout the answer, the candidate maintains a clear structure and the vast use of examples helps to add detail. Reference is made to a wide range of measures, which avoids repetition and makes it easier for the candidate to pick up marks. Overall this is an excellent answer and now well worth full marks.

Answer B

(a) Methods of Combat:

- *Mosquito nets – People wear these at night to avoid getting bitten.*
- *Coconuts – These are injected with bacteria that harms the mosquito.*
- *Spraying DDT – This is harmful to the mosquitoes and kills them.*
- *Draining irrigation channels – This takes away the water in which the mosquitoes breed.*
- *Education – People can be educated on the dangers of malaria.*

(2 marks)

(b) There are many methods used to combat malaria, with some being more effective than others. Sometimes these methods are very expensive and people cannot afford them. Some drugs that are available are not accessible to those who need it most, as they are too expensive. Bill Gates has set up a foundation to help people with malaria but again, not all communities are feeling the benefits.

Irrigation channels can be drained and emptied on a regular basis but this is expensive and very wasteful, particularly in areas that suffer from a lack of rainfall. **(1 mark)** *Mosquito nets can get damaged and get holes in them and are expensive to replace. If the holes are big enough to allow mosquitoes through then it defeats the purpose of using them and the disease continues to spread.* **(1 mark)**

4 out of 10 marks

What makes this a weak answer?

Given that the candidate has answered part (a) with simple descriptive points they can achieve no more than **2 marks** for this part of the question. This is an example of very poor exam technique, particularly at Higher level. The candidate clearly demonstrates they know what methods are involved. However, they don't explain how each method actually helps to control the disease, therefore failing to answer the question.

The second part of the answer is over-generalised with poor use of examples. Although the candidate has chosen to answer parts (a) and (b) separately, you should still see a connection between the two sections. Here, new methods have been introduced in part (b) that were not discussed in part (a) and as a result have not been explained fully, e.g. the Bill Gates Foundation. This is a good example to include. However, this point, like many others, has not been developed, proving costly to the overall marks awarded.

Altogether, the candidate clearly has some knowledge of the methods used to combat the disease. Yet, poor exam technique has prevented this candidate from answering the question fully, resulting in an answer that only achieves **4 out of 10**.

A summary of all the methods used to combat malaria can be found below. Familiarise yourself with these and make sure you are able to explain them.

> ### Top Tip ✔
>
> A simple list of factors rather than fully developed, explanatory points will prove detrimental to the overall number of marks awarded. For example, for a **6 mark** question, a maximum of **2 marks** can be awarded for answers that are purely descriptive.

Figure 10.4: Methods to combat malaria

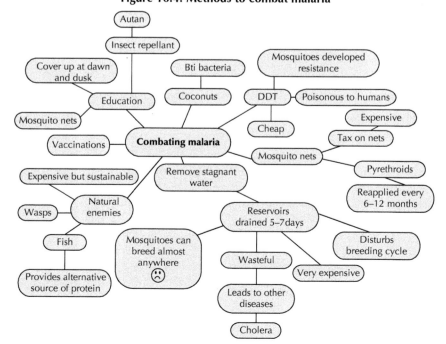

Maximising Marks

For one mark you should give one detailed explanation, e.g.

'Draining areas of **stagnant water,** such as puddles and swamps every **7 days**, can reduce the number of **breeding grounds** for the mosquitoes to **lay their eggs.'** (**1 mark**)

OR give a limited description/explanation of two factors, e.g.

'Once people had been contaminated with malaria, attempts were made to cure them by **killing the plasmodium parasite.** Drugs like **Quinine, Chloroquine** and **Malarone** were all developed in an attempt to kill the parasite.' (**1 mark**)

Primary Health Care

PHC is the provision of basic, low cost health services that do not necessarily require fully trained medical staff. You should be able to explain the strategies involved and how they are suited to developing countries.

Figure 10.5: Primary Health Care strategies

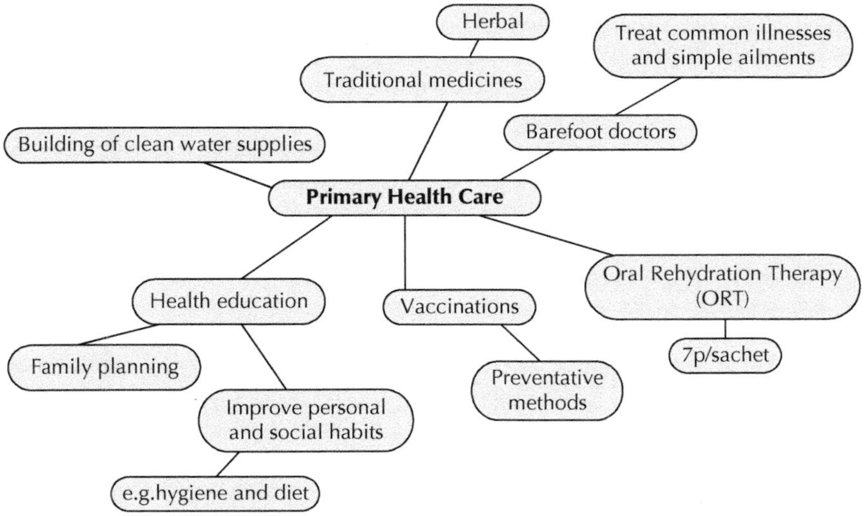

Q **Explain** why Primary Health Care strategies are suited to developing countries.

5 marks

Answer A

Primary Health Care strategies are good for developing countries as they can improve communities without the need for lots of money. Barefoot doctors are people who are trained in basic medical care. They tour around local villages and help people when they are sick. They are trained to treat simple illnesses and will move around a lot visiting different villages on different days. **(1 mark)** ORT, vaccination programmes and education schemes are other examples. These are good for local communities as it means they are being educated on how to keep themselves healthy. These methods also help to maintain a healthy community and reduce the rate of infant mortality. **(1 mark)** Mosquito nets are also given out to help reduce the number of people getting the disease. This preventative method is very effective as it saves money on treating the disease if it can be prevented in the first place. **(1 mark)**

3 out of 5 marks

What makes this a weak answer?

Having achieved 3 out of 5, this is not a bad answer. The candidate has given some examples of Primary Health Care strategies and linked them back to the question. However, it is a lack of detail that has cost this candidate marks. Although Oral Rehydration Therapy, vaccinations and education programs were mentioned, the candidate would have received more marks if they had taken each of these factors in turn and discussed them separately. Therefore a better structure with additional detail would help this candidate earn more marks.

Answer B

Barefoot doctors are local people who are trusted within communities. They carry out treatment for common illnesses and simple ailments, preventing sick people from travelling far to a medical centre. **(1 mark)** Barefoot doctors will use cheaper, traditional remedies as opposed to expensive cures given at the hospital. **(1 mark)** Local initiatives may also be supported by small local health centres staffed by doctors, with only serious cases going to hospital. This reduces the burden on hospital care. **(1 mark)**

Oral Rehydration Therapy (ORT) is used to tackle diarrhoea and dehydration. ORT helps to raise the survival rates of babies and young children, helping to reduce rates of infant mortality. **(1 mark)** Clean water supplies and decent sanitation may be installed in villages, often with community participation. **(1 mark)**

Vaccination programmes against diseases such as polio, measles and cholera are now widely used in some areas. Such programmes focus on preventative methods as opposed to more expensive curative medicines. **(1 mark)** In addition, Health Education schemes are being introduced in schools and communities to target women and children in relation to hygiene and diet. **(1 mark)** If communities can improve the health of their population, life expectancy will improve. They will have a healthier workforce who can help develop the country and infant mortality will decrease allowing more children to go to school and creating a skilled workforce. **(1 mark)**

AIDS is a major problem in areas such as Malawi. Medecins Sans Frontieres (MSF) have been working in Southern Malawi since 1999. Here, 25% of the adult population are HIV positive. They opened a clinic where sufferers can go for treatment. The clinic also offers advice on safe sex, *contraception and offers counselling. In 2001, a hospital and community health centre were opened, which meant more people could be treated effectively.* **(1 mark)**

5 out of 5 marks

What makes this a strong answer?

Unlike in answer A, this candidate has taken each strategy in turn and produced an answer that is detailed, developed and well structured. Each point has been explained fully and examples have been used to further strengthen the answer.

Maximising Marks

Maximising marks using case studies

Learning your case studies is an important part of your exam preparation. Being able to give specific named examples adds detail to your answer and is a quick and easy way of picking up extra marks. At Higher level, it is not enough to simply name the area you have studied – you must demonstrate your knowledge to the examiner by referring to **specific** examples throughout your answer.

For two specific named examples within the case study area you will be awarded one mark. A maximum of two marks can be gained for any one question.

Glossary

Birth rate: The number of babies born per 1000 women per year.

Death rate: The number of people who die per 1000 per year.

Development: The progress of an advanced country in terms of wealth, health and technology.

Economic indicator: Used to assess how economically developed a country is, e.g. GDP.

Export: Goods sold to other countries.

GDP per capita: Gross Domestic Product is the total value of goods and services produced in a year within a country. 'Per capita' means 'per head': GDP divided by population.

Import: Goods bought by a country from abroad.

Infant mortality rate: The number of infants dying under one year of age, per 1000 live births.

Social indicator: A feature of society that can be measured to show a country's level of development, e.g. death rate.

Section 11: Climate Change

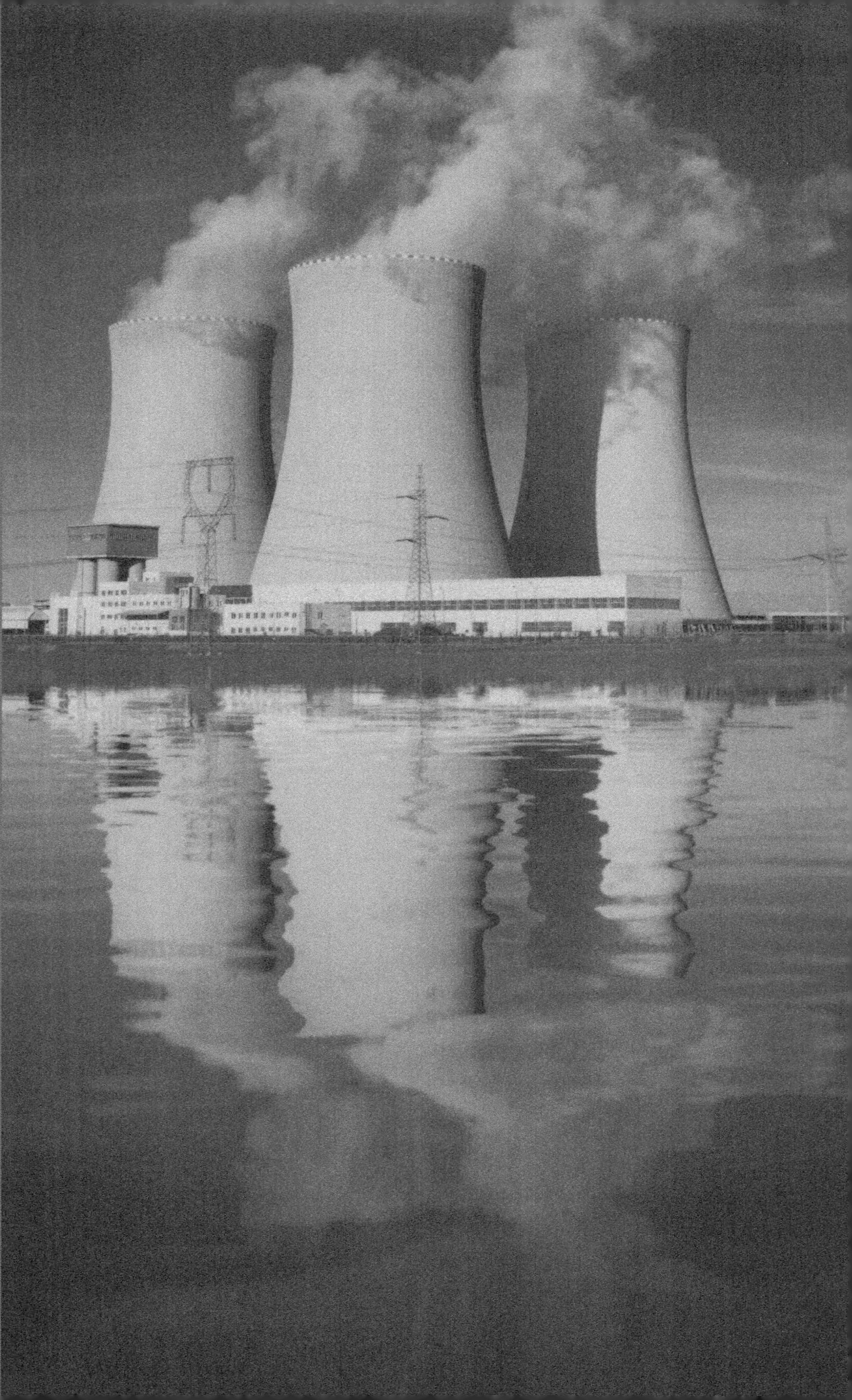

Climate Change

SECTION
11

Topics include:

• Physical and human causes of climate change

• Local and global consequences of climate change

• Management strategies and their limitations

Human and physical causes

You may be asked to explain the human or physical causes of climate change or a combination of both. A summary of these factors is shown below.

Figure 11.1: Human causes of climate change

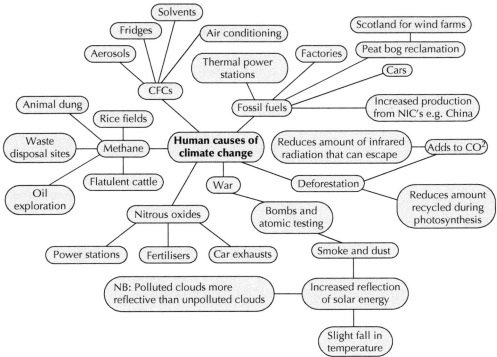

Figure 11.2: Physical causes of climate change

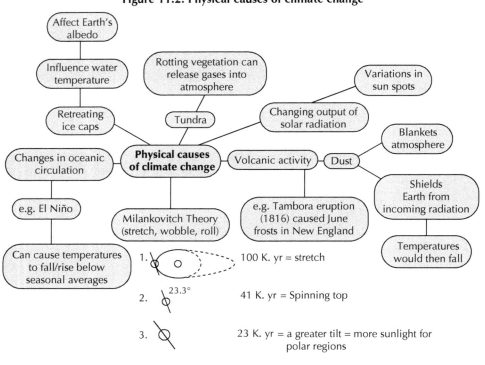

Affect Earth's albedo

Influence water temperature

Rotting vegetation can release gases into atmosphere

Variations in sun spots

Retreating ice caps

Tundra

Changing output of solar radiation

Blankets atmosphere

Changes in oceanic circulation

Physical causes of climate change

Volcanic activity

Dust

e.g. El Niño

Milankovitch Theory (stretch, wobble, roll)

e.g. Tambora eruption (1816) caused June frosts in New England

Shields Earth from incoming radiation

Can cause temperatures to fall/rise below seasonal averages

Temperatures would then fall

1. 100 K. yr = stretch

2. 23.3° 41 K. yr = Spinning top

3. 23 K. yr = a greater tilt = more sunlight for polar regions

---

Q **Explain** the human factors that have contributed to changes in global temperatures.

5 marks

Look at the two answers below. What differences do you notice between them? Think about the use of examples, level of detail and relevance to the question.

Answer A

In recent years, human activity is seen as the most important factor in global temperature change, particularly through the increased consumption of fossil fuels. Thermal power stations, heating systems, deforestation and road vehicles all contribute to increasing global CO2 levels. **(1 mark)**
Peat bog reclamation is another cause, for example, in Scotland where land is being developed for wind farms. **(1 mark)** *There is also the added problem of increased production from newly industrialised countries such as China. CO2 from combustion is responsible for 55% of greenhouse gas from human activity.* **(1 mark)**

Top Tip

For every factor, include an explanation for **why** it is contributing to a change in temperature, e.g. methane is released from cattle's digestive system and beef is increasingly in demand across the world.

Increased output of nitrous oxides due to vehicle exhausts, increased use of fertilisers and power stations are all a result of an increasing population with high demands. **(1 mark)**

CFCs are a very powerful greenhouse gas released from disused fridges when the foam insulation inside them is shredded. **(1 mark)** *The coolants used in fridges and air conditioning systems create CFCs, which are generally safe but can be harmful when released if appliances are not disposed of correctly.* **(1 mark)**

The rapidly increasing populations in Asian countries such as China and India have resulted in an increase in methane production, particularly through rice paddies and belching cows, as more food is being produced to feed the population. **(1 mark)** *Methane is also released from landfill sites, oil exploration and permafrost melting in Arctic areas.* **(1 mark)**

5 out of 5 marks

Answer B

A number of human factors have been involved in terms of increasing global temperatures. Deforestation, power plants and cars all contribute to increased levels of harmful gases. These human activities all contribute to the release of gases such as CFCs, methane and nitrous oxides. **(1 mark)**

The use and testing of atomic bombs during the 1940s, 50s and 60s released radioactive particles into the atmosphere and beneath some oceans. **(1 mark)** *Furthermore, an increase in volcanic eruptions and the consumption of meat are also major factors in contributing to global climatic change.*

2 out of 5 marks

What answer comes out on top?

Answer A is well structured and is of a high standard. Unlike answer B, this candidate has taken each human factor in turn and explained how it is contributing to changes in temperature. They have included examples from the local level such as the use of car exhausts and air conditioning all the way up to global activities such as deforestation and increased industrial production.

On the other hand, answer B achieves 2 out of 5. The first sentence is a simple rewording of the question and achieves no marks. Everything you write on your exam paper should be there for a reason – simply rewording the question like this is a waste of time. The candidate has explained some human factors. However, on the whole this answer is largely descriptive and does not contain enough detail to achieve full marks. Physical factors have also been included, e.g. 'an increase in volcanic eruptions', which is irrelevant and gains no credit. Overall this is a vague response that fails to fully answer the question.

Maximising Marks

In a question that asks you to **'explain factors'**, you should take each factor in turn. State the factor **then** explain it, before moving on to your next point. This way you can ensure that each point is developed whilst linking back to the question, e.g.

'**CFCs** are a very **powerful greenhouse gas** released from **disused refrigerators** when the **foam insulation** inside them is **shredded.' (1 mark)** rather than, 'Deforestation, power plants and cars all contribute to increasing the level of harmful gases.'

At Higher level, answers that consist of lists are heavily penalised. Practicing how to structure your answers will help you avoid this mistake.

Local and global effects

> **Q** **Discuss** the possible impacts of global warming from a **local** to **global** scale.
> **6 marks**

Before you start, identify the key words in this question. If you do not refer to effects at **all levels** you cannot achieve full marks. A possible answer is shown below.

Precipitation is likely to increase on a global scale, particularly in the winter in northern areas such as Scotland. However, some areas like the Great Plains, USA, may experience drier conditions. **(1 mark)**

Due to **increased temperatures,** there may be **longer growing seasons,** particularly in areas of northern Europe. This will lead to **increased food production** and a **greater variety of crops. (1 mark)** However, increased temperatures may also lead to an **increase in tropical diseases** such as malaria, with an estimated 40 million more people being at risk of contracting the disease. **(1 mark)**

As it becomes warmer, sea-ice and glaciers in areas like Greenland and Antarctica will melt, causing **sea-levels to rise.** This is a major threat to communities living in low-lying coastal areas (e.g. the Maldives). **(1 mark)** Already in areas like Bangladesh, large numbers of

people have been displaced *and* **farmland/property has been destroyed** *due to an* **increase in floods. (1 mark)**

More extreme and variable weather *is becoming much more frequent around the globe. Events such as Hurricane Katrina and droughts in Ethiopia are becoming more intense, greatly affecting local communities.* **(1 mark)**

In addition, changes to **atmospheric patterns** *and* **ocean current** *circulation will have a large impact, for example in the Atlantic, the Thermohaline circulation will start to lose impact on north-western Europe, resulting in winters that will be considerably colder than before.* **(1 mark)**

Lastly, global warming will have a dramatic impact on **wildlife** *across the globe. It is understood that there will be extinction of at least 10% of land species, while coral reefs are likely to become depleted.* **(1 mark)**

What makes this a strong answer?

Structure
This answer has been broken down, with the candidate taking a new paragraph for each new point. This not only makes it easier to see what you have already included (avoiding repetition) but it also makes it easier for the examiner as they are not having to trawl through large paragraphs of information in search of a mark.

Variety
This answer includes a whole range of impacts at all levels across the globe (illustrated in bold). The candidate has discussed many of them at a global level before discussing their impact locally.

Examples
Specific named examples are a good way of showing the examiner that you know what you are talking about. One mark is awarded for reference to **two** specific named examples within your case study.

Management strategies and their limitations

You should be aware of some of the methods employed in an attempt to reduce the effects of climate change. Figure 11.3 illustrates some of the solutions with which you should be familiar.

Figure 11.3: Solutions to climate change

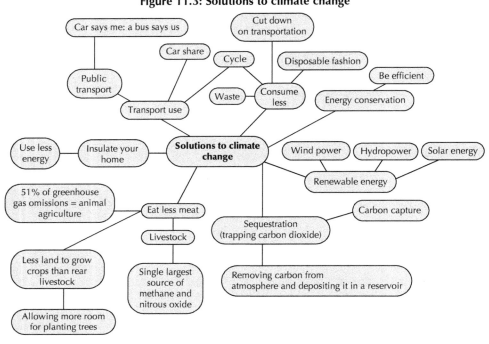

> **Q** **Discuss** ways in which people can try to reduce the effects of climate change. **6 marks**

Answer A

Using **renewable energy sources** like wind energy, solar energy, hydroelectric power and geothermal energy reduces the reliance on and usage of fossil fuels. **(1 mark)** Using alternative greener sources of energy does not produce harmful emissions and therefore reduces the levels of carbon dioxide in the atmosphere. **(1 mark)**

People are encouraged to reduce their reliance on cars by using **public transport, car-pooling or cycling**. Schemes like the **'cycle to work scheme'** have been launched by local councils to encourage more people to use 'greener' modes of transport. **(1 mark)** In the USA, 'car-poolers' are able to use the fast lane during peak times while single drivers are fined if seen to be using this lane. **(1 mark)** Cars are now being designed to be more eco-friendly and fuel-efficient in an attempt to reduce the levels of carbon dioxide released. **(1 mark)**

Consumers are being encouraged to **consume less** in an attempt to conserve resources and energy. Pollution is produced at each stage of the consumerist process and therefore, by reducing, reusing and recycling, greenhouse gas emissions will be reduced. **(1 mark)** Shops have now started charging people for carrier bags. This encourages people to re-use their own plastic bags, helping to minimise waste. **(1 mark)**

Organic farming *and eating less meat will further reduce the levels of methane and nitrous oxides in the atmosphere.* **(1 mark)** *Education campaigns have also been used to inform people of the importance of protecting the Earth and to encourage people to be more eco-friendly, helping to change peoples' attitudes and actions.* **(1 mark)**

6 out of 6 marks

What makes this a strong answer?

With 6 out of 6, this candidate fully addresses the question and considers a range of different ways people could try to reduce the effects of climate change. Each point has been well developed and the answer has a clear structure, limiting any repetition. Given this is a '**discuss**' question, the candidate has successfully considered the different methods involved before commenting on the effectiveness. This is a very detailed response and one worthy of full marks.

Answer B

People can try to reduce the effects of climate change by adopting greener practices. Switching lights off and taking the bus will help to reduce levels of greenhouse gases on an individual basis. Education campaigns have been launched to inform people of these simple 'greener' actions. **(1 mark)**

Recycling is also effective. Local councils like Glasgow City Council have given each household a recycling bin. People can drive to work with their friends or take the bus. People can also grow their own food. This cuts down on food miles and food packaging, reducing greenhouse gas emissions. **(1 mark)** *Natural fertilisers can be used, such as salt, reducing the reliance on harmful chemical fertilisers. This is a lot more environmentally friendly and dramatically cuts down on pollution.* **(1 mark)**

3 out of 6 marks

What makes this a weak answer?

This is a fairly basic answer, with the candidate gaining 1 mark in the first paragraph for a limited description of two factors. The second paragraph has good intentions with reference to recycling. Yet, this point has not been developed, failing to pick up any marks. The last paragraph is slightly better with two marks being awarded. However, on the whole, the candidate has not fully explored the issue, failing to discuss a range of topics in detail. Opportunities to gain marks have been lost due to a lack of explanation and poor exam technique, resulting in a total of 3 out of 6 marks.

Q **Evaluate** the effectiveness of the strategies used to manage the local effects of climate change. **5 marks**

Look at the two answers below. What differences do you notice between them? Think about level of detail and relevance to the question.

Answer A

As temperatures rise, the effects of climate change are increasingly felt in local communities. Flooding becomes a major issue, particularly around low-lying coastal areas. This can wipe out vast quantities of crops and agricultural products, besides ruining people's homes. Many strategies have been put in place in order to manage flooding such as flood defences, afforestation and building settlements on higher land.

Climate change can cause a threat to water security and a threat to local ecosystems. This can put people at risk and can lead to a reduction in wildlife habitats. Drought is also a problem that affects people at the local level. 2004 to 2006 was one of the driest periods on record in the UK. This placed huge pressures on the water system and as a result, many restrictions were put in place. Rivers dried up, groundwater levels fell to their lowest on record and fish stocks dwindled. This had major implications for the people of the UK.

0 out of 5 marks

Answer B

To manage the effects of extreme weather events, flood defences have been built, for example the Thames Barrier. This has protected London from flooding on many occasions and is predicted to provide protection for many years to come. **(1 mark)** *However, in approximately 50 years the original barrier may need to be strengthened or a second barrier built in order to maintain its effectiveness.* **(1 mark)** *In addition, advance warning systems should be developed to prepare householders for the potential risk of flooding. This would enable people to plan ahead, which limits levels of devastation.* **(1 mark)**

Drought is increasingly becoming more of a threat as a result of increasing temperatures bringing long dry spells of weather. This has been managed through the implementation of hosepipe bans in many parts of England. However, these bans are very unpopular and prove difficult to enforce. **(1 mark)** *Another strategy employed was the development of a desalinisation plant in London. This is very expensive to operate and as a result is used only in periods of extreme drought.* **(1 mark)**

5 out of 5 marks

What answer comes out on top?

Answer A is a very basic response that fails to refer to the question. Although the candidate has given valid information with regard to the local effects of climate change, they have not discussed the management strategies involved and therefore have failed to answer the question. When asked to **evaluate,** you are being asked to make a judgment of the success, failure or impact of something. As the candidate has not done this, marks cannot be awarded.

Answer B, on the other hand, achieves 5 out of 5 through the inclusion of detailed, evaluative points. Although the candidate has stated the strategies employed, the marks have been awarded where they have evaluated the success of each strategy. The decision to split the answer into separate paragraphs ensures a clear structure and helps to avoid repetition. This is a well-planned answer and one worthy of full marks.

Glossary

Chlorofluorocarbon (CFC): An organic compound responsible for global warming.

Fossil fuels: Energy resources such as natural gas, oil and coal formed from the fossilised remains of plants and animals.

Global warming: The gradual rise in world temperatures due to the increase of greenhouse gases as a result of human activities.

Greenhouse effect: A natural process by which the atmosphere traps some of the sun's energy.

Ozone layer: A layer of gas in the atmosphere that protects the Earth from harmful radiation from the sun.

Section 12: Trade, Aid and Geopolitics

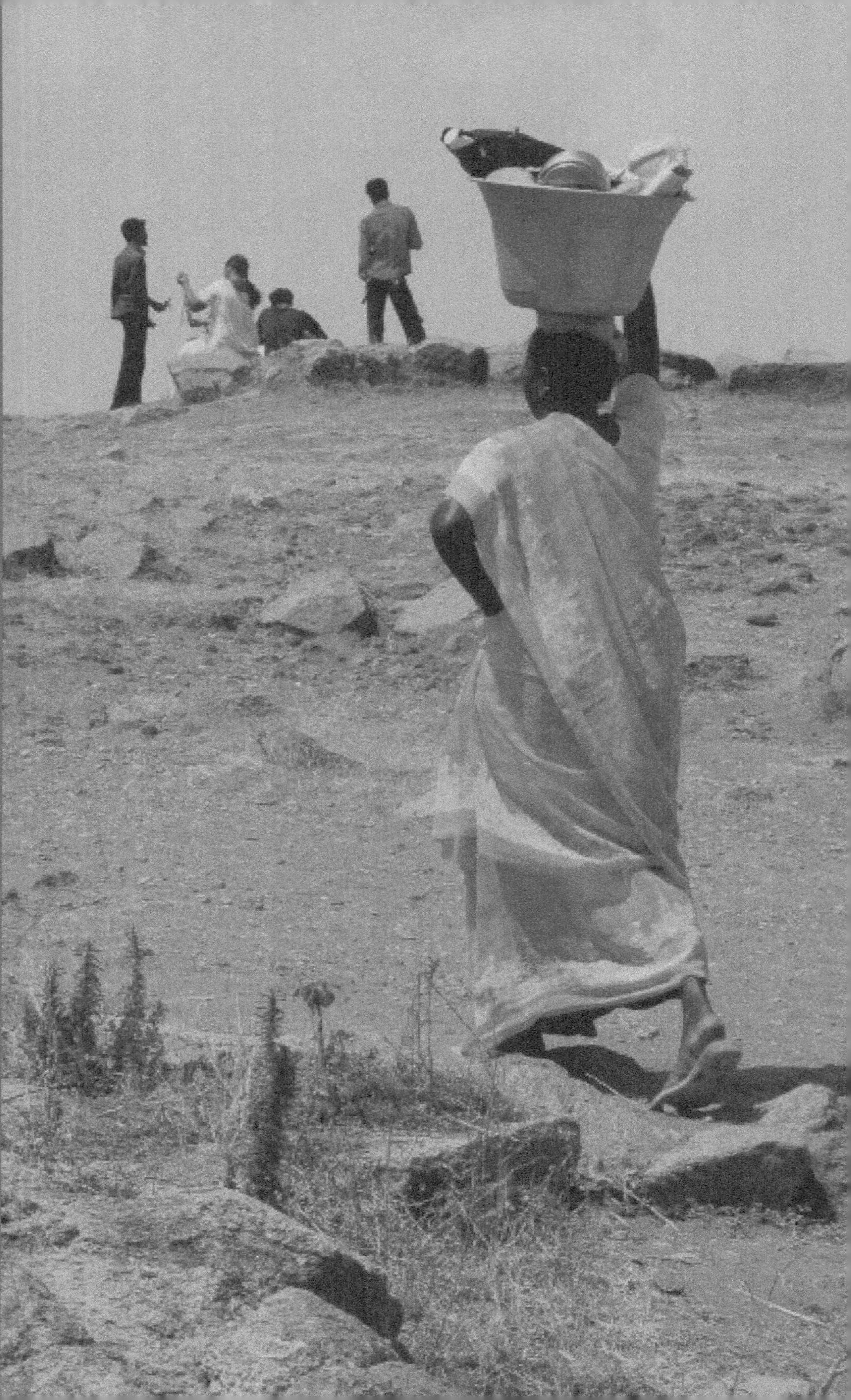

Trade, Aid and Geopolitics

Topics include:

- World trade patterns
- Causes of inequalities in trade
- Impact of world trade patterns
- Aid and other strategies to reduce trade inequalities and their impact

Causes of inequalities in trade

With more than half of world trade taking place between only 8 countries, the trading process is very unequal. Figure 12.1 illustrates some of the reasons behind these inequalities with which you should be familiar.

Figure 12.1: Inequalities in world trade

Difficult to escape poverty

Unstable

Difficult to plan improvements

Little purchasing power

Reliance on primary products

Developing countries

Little manufacturing industry

No powerful trading groups

Inequalities in world trade

Forcible opening of markets

Dumping

Products exported at price below production cost

World bank

International trade agreements

EU and US subsidise products

International Monetary Fund

119

Q Study Figure 12.2. **Account for** the inequalities in trade shown in the table below.

5 marks

Figure 12.2: Inequalities in trade

		Germany	Thailand	Kenya
Trade statistics (million US$, 2012)	Exports	1,492,000	226,200	5942
	Imports	1,276,000	217,800	14,390
	Balance of trade	216,000	8400	–8448
Economic Indicators (2012)	GDP per capita (US$) (PPP*)	39,100	1800	10,300
	% employed in agriculture	2%	38%	75%
	% employed in manufacturing	24%	14%	10%
	% employed in services	74%	48%	5%
	Main exports	Motor vehicles, machinery, chemicals, computer and electronic products	Electronic, computer parts, automobiles and parts, electrical appliances, machinery and equipment textiles	Tea, horticultural products, coffee, petroleum products, fish

PPP* = purchasing power parity

Answer A

Developing countries like Kenya often rely on primary products for trade. However, the price of primary products fluctuates on the world market, therefore profits are limited, with Kenya earning only $5942 from exports in 2012. **(1 mark)** *Similar raw materials are often abundant* in many other countries, which keeps competition high and prices low. **(1 mark)** *Developed countries, on the other hand, export valuable manufactured products like cars and electronics, which adds value and provides increased profits.* **(1 mark)** *Developed countries also impose tariffs and quotas on imports, which usually works in their favour, e.g. Kenya's export of coffee to the European Union is subject to a tariff of 9%, while other countries are subject to a 3.1% tariff.* **(1 mark)**

In addition, developing countries are often very dependent on one or two products, e.g. sugar or tea. **(1 mark)** *However, the price for these products is unstable and is often determined by developed countries through trading on 'commodity exchanges' around the world, e.g. the New York Mercantile Exchange.* **(1 mark)** *If the price or demand falls, the country's income can be affected, limiting their ability to buy imports.* **(1 mark)**

5 out of 5 marks

What makes this a strong answer?

The candidate has used the table well to explain inequalities in world trade patterns. They have provided a number of reasons and extracted information from the source to back up each point. As you can see here, the candidate has not relied too heavily on describing the table but instead they have used the information to support their response. This is a good example of how to use a source to help develop your answer and includes more than enough detail to earn full marks.

Answer B

In 2012, Germany earned $1,492,000 through exporting products, whereas Kenya earned considerably less with $5942. Kenya spends a lot more on imports than it earns through exports, resulting in a negative trade balance of $8448. **(1 mark)** *The majority of Kenya's population are employed in agriculture (75%), whereas only 2% of Germany's population work in this industry. The agricultural industry can be unpredictable and brings in little money for the country.* **(1 mark)** *Thailand on the other hand is more balanced with 38% employed in agriculture and 48% employed in the service industry. Both Thailand and Germany export a lot of electronic products and machinery whereas Kenya relies on exporting primary products such as tea and coffee, which can result in an unstable income.* **(1 mark)**

3 out of 5 marks

What makes this a weak answer?

Having misread the question, the candidate solely describes the table, giving no explanation for the inequalities in trade between countries. The candidate has referred to most of the development indicators given yet has failed to link this information to the question. However, given the candidate has described the table well with limited explanation, a maximum of 3 marks can be awarded for description.

Maximising Marks

A maximum of three marks will be awarded for answers that are purely descriptive. Therefore, in order to achieve full marks you need to **use** the information provided to help develop your answer and **explain** each point, e.g.

'Developing countries like Kenya often rely on primary products for trade therefore **profits are limited,** *earning only $5942 from exports in 2012.'* **(1 mark)**

Impact of strategies to reduce inequalities in world trade

Many strategies have been put in place in an attempt to reduce the inequalities in trade between countries. Some of the strategies employed are shown in Figure 12.3.

Figure 12.3: Strategies to reduce inequalities in world trade

Q	**Explain** the strategies used to reduce the inequalities in world trade. **5 marks**

Answer A

The Fair Trade initiative has been introduced to help developing countries earn a fair price for their products. It means that regardless of the market value, farmers and producers are guaranteed a fair price to cover the cost of production. **(1 mark)**

Consumers in developed countries are then encouraged to buy products with the Fair Trade logo. This means that the farmer can plan ahead as they will now be getting an income that will be good for their family.

There is also the World Trade Organisation (WTO), which was established in 1996. This was set up to foresee exchanges in trade around the world. The WTO promotes free trade and the removal of tariffs and quotas. They aim to settle any disagreements in trade between different countries. **(1 mark)**

2 out of 5 marks

> **Top Tip** ✔
>
> Scribbling down key words in a spider diagram before attempting a question like this will help you organise your answer and helps to ensure you don't miss anything out.

> **Top Tip** ✔
>
> In order to explain the strategies used, you should **explain** the **effectiveness** of them with regard to reducing inequalities in world trade.

What makes this a weak answer?

This answer starts off well and includes good points about Fair Trade and the WTO. However, in the second paragraph the candidate has started to waffle – losing focus from the question. This has earned them no additional marks and therefore has wasted time.

Furthermore, this answer is made up of three short paragraphs covering only **two** strategies. There is not enough in this answer to qualify for five marks. It is important to always look at the number of marks awarded and write your answer accordingly. For one mark you should give a detailed explanation or a limited description/explanation of two factors.

> **Top Tip** ✔
>
> It is a good idea to stop half way through your answer and re-read the question. This will allow you to make sure you are on track and that your information is relevant.

Answer B

One strategy is to remove trade barriers. This enables developing countries to have access to profitable markets in the developed world. **(1 mark)** *Also, some countries, e.g. in the Caribbean, are attempting to diversify their trade by developing non-traditional exports such as new crops, while others are pursuing new markets.* **(1 mark)**

Another strategy is the creation of trade alliances, e.g. the Caribbean Community and Common Market (CARICOM), which was established to promote trade between Caribbean countries. **(1 mark)** *This resulted in customs duties between member states being removed, allowing even the smallest countries to have access to a regional market, which reduced their costs.* **(1 mark)** *Member countries are encouraged to purchase raw materials from other CARICOM countries. This has shared the benefits of industrialisation and has encouraged industries to set up in smaller countries.* **(1 mark)** *This not only generates more trade and wealth, but also creates employment, raising the GDP.* **(1 mark)**

Fair Trade aims to guarantee a fair price for produce by covering the cost of production regardless of fluctuations in the market price. **(1 mark)**

In addition, five-year rolling contracts can be given, resulting in a secure income. This allows long-term planning to take place for investment in farm machinery, education and health services, enabling communities to plan for the future and aiding development. **(1 mark)**

5 out of 5 marks

What makes this a strong answer?

The use of examples in this answer adds depth and demonstrates a good understanding, resulting in a response that is detailed and well developed. The candidate has covered a number of different strategies and explained how effective they are with regard to reducing inequalities in world trade. Taking a new paragraph for each strategy helps to split up the answer and maintain a strong structure. Overall the candidate demonstrates good exam technique and full marks are well deserved.

Glossary

Exports: Goods sold to other countries.

Fair Trade: When the producer is guaranteed a fair price for their goods.

Quota: A limit imposed on the amount of goods imported.

Tariff: A tax (charge) imposed on imports or exports.

Trade: The movement of goods and services between countries.

Trade alliance: A group of countries that have joined together to give themselves more economic power, e.g. the European Union (EU).

Trade balance: The difference between value of imports and value of exports.

Trade deficit: When the value of a country's imports exceeds the value of its exports.

Trade surplus: When the value of a country's exports exceeds the value of its imports.

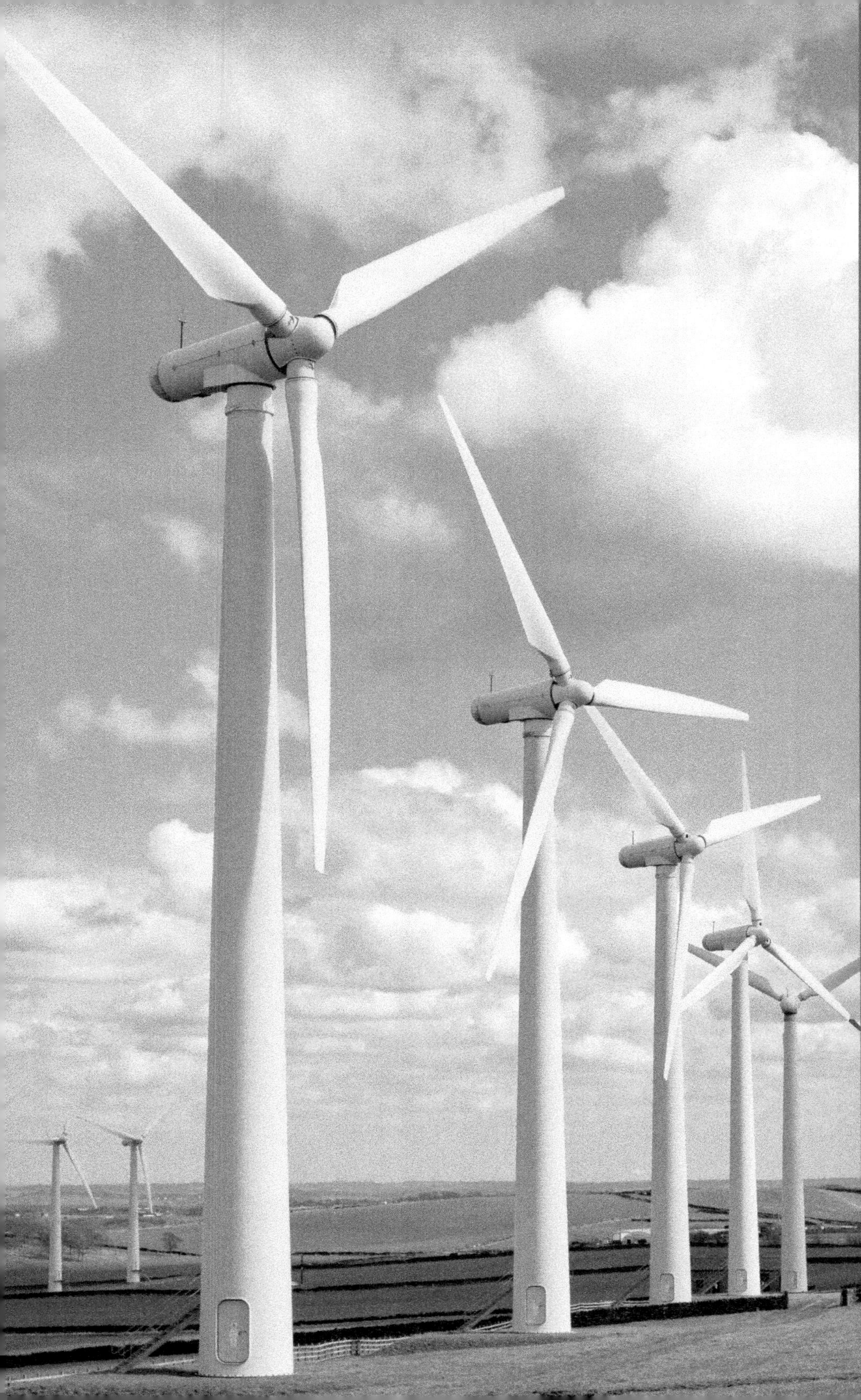

Section 13: Energy

Energy

Topics include:

- Global distribution of energy resources
- Changing levels of demand for energy in developed and developing countries
- Effectiveness of renewable and non-renewable approaches
- Suitability of renewable energy approaches within different countries

Changing levels of demand for energy

Q Study Figure 13.1. **Explain** the differences in energy consumption between developed and developing countries.   **5 marks**

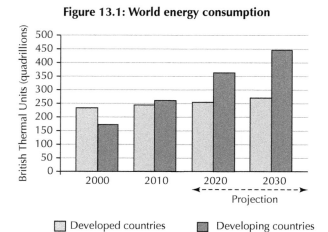

Figure 13.1: World energy consumption

Maximising Marks

When asked to explain the differences shown on a graph, you must first work out what the graph is telling you. No marks will be awarded here for describing the differences – you must **account** for the differences you see.

From looking at the graph, the amount of energy used by 'developed countries' in 2000 is slightly higher than the amount used by 'developing countries'. However, from 2010–2030 this is forecast to change. The energy use in 'developing countries' is expected to increase at a much faster rate, whereas use in 'developed countries' is expected to rise very slowly. Your task, therefore, is to explain **why** this is the case.

Answer A

As shown on the graph, energy levels in 'developing' countries are increasing dramatically whereas in 'developed' countries, they are staying the same. In 2000, energy use in 'developing' countries was around 170 btu whilst in 'developed' countries it was 240 btu. In 2010, the energy usage in 'developing' countries increased to 260 btu and is expected to reach 360 btu by 2020 and 450 btu by 2030. The projected energy use for 'developed' countries in 2030 is only around 270 btu.

The energy use in 'developing' countries is growing so rapidly due to an increase in wealth. People all over the world are becoming richer and so technology is no longer confined to the 'developed' world. More and more people now have access to lights, electricity and central heating. **(1 mark)** *Globalisation has a major role to play in this increased use of energy. People are also relying more on transport to get around and so car ownership has increased dramatically.*

1 out of 5 marks

What makes this a weak answer?

Although this answer is well structured, the information given is basic and largely irrelevant. In the first paragraph the candidate has described the differences, making no attempt to explain them. This wastes time and earns no marks.

The second paragraph starts well, explaining why energy use in 'developing' countries may have increased. One mark has been earned here for a limited explanation of two points. The candidate then goes on to mention globalisation. This point has potential to earn additional marks. However, the point is left undeveloped and no explanation is added. The candidate also mentions transport use, yet, again, fails to develop this point, missing the opportunity for marks to be awarded.

Answer B

In 'developed' countries population growth rates are stable and in some places are even declining. Therefore, there is no increase in demand for energy. **(1 mark)** *Companies are also looking to make new products more energy-efficient while individuals are being encouraged to be more 'eco-friendly', helping to stabilise or reduce energy consumption.* **(1 mark)**

However, it is in 'developing' countries where populations are increasing rapidly that communities are becoming richer. **(1 mark)** *With increased wealth comes an increase in standards of living. Billions of people now have access to electricity, which they didn't have before. This causes an increase in the use of lighting and appliances such as televisions, washing machines, computers and air conditioning, which will all cause energy usage to increase.* **(1 mark)**

This economic growth in 'developing' countries is linked to 'energy-hungry' manufacturing industries. As more and more companies set up and relocate to 'developing' countries, their use of energy increases. Goods produced by these industries are often sold to 'developed' countries, which are then transported around the world using even more energy. **(1 mark)**

In addition, as people in 'developing' countries become richer, car ownership will increase as well as the development of public transport systems. This increase in the use of transport will further increase energy levels. **(1 mark)**

5 out of 5 marks

What makes this a strong answer?

Through analysis of the graph, the candidate has identified that energy use in 'developing' countries is forecast to increase at a much faster rate than in 'developed' countries. They have then proceeded to explain why. By referring to 'developed' countries **then** 'developing' countries, the candidate has created a clear and precise structure that fully addresses the question. By making good use of the resource provided, a detailed response has been created, typical of an A student.

Suitability of renewable energy approaches

> Q Referring to different countries, **evaluate** the suitability of renewable approaches to generating energy. **6 marks**

Take a moment to jot down all the approaches you have studied in class. This will help add structure to your answer, particularly if you are feeling the constraints of time.

An example of what you should include is summarised below.

Top Tip

When asked to **evaluate** you are expected to make a judgement on the success, failure or impact of something. In this case, you are being asked to **evaluate** the effectiveness of renewable energy approaches.

Figure 13.2: Renewable energy approaches

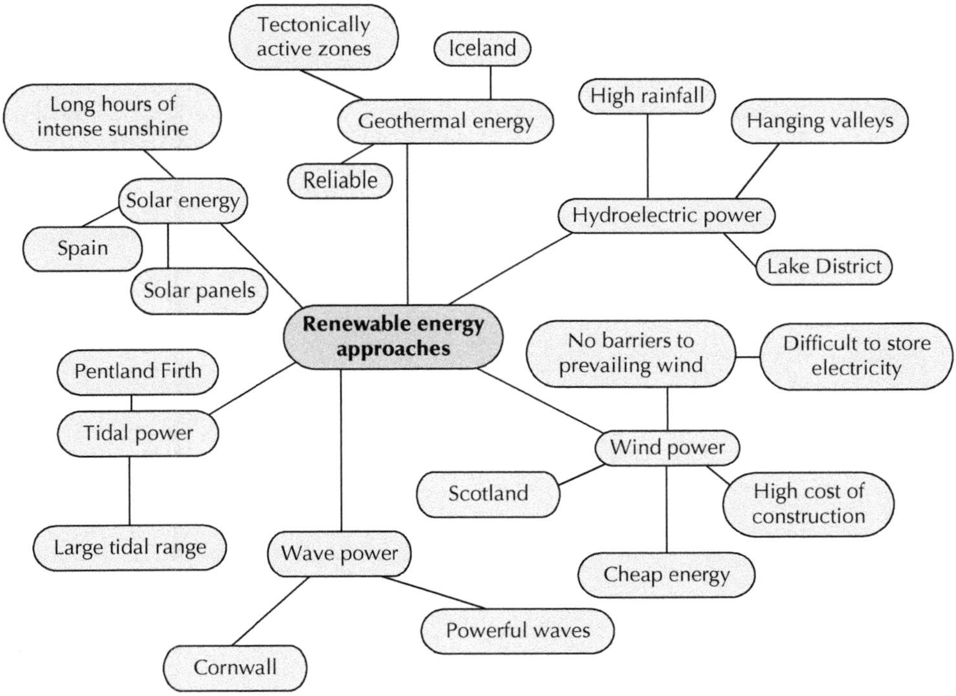

Look at the answer below. What is good about this answer?

Solar energy is most effective in places where there are long hours of intense sunshine to power the solar panels, e.g. in Spain **(1 mark)**. *Countries such as Scotland are an ideal location for hydroelectric power as it is most effective in areas where there is high rainfall to ensure reservoirs are always at maximum capacity.* **(1 mark)** *Hanging valleys, which are abundant in Scotland, are an ideal site for hydroelectric power as they allow the vertical drop of water needed to power the turbines.* **(1 mark)**

Wave power is very effective in the south of England, such as in Cornwall where the fetch is large enough to give power to the waves. Tidal power is also effective here given the large tidal range available to turn turbines. **(1 mark)**

Geothermal energy is common in places like Iceland where there is tectonic activity. The magma creates a heat source close to the Earth's surface, which can be used to generate steam. **(1 mark)**

Wind power is effective in landscapes that are fully exposed to the wind, allowing for a regular and reliable source of air to turn the turbines, e.g. in Germany. **(1 mark)** *However, it is difficult to store the energy produced from turbines and so there is concern about how to bridge the energy gap when the turbines are not generating electricity on calm days.* **(1 mark)**

6 out of 6 marks

What makes this a strong answer?

Structure
The candidate has taken a new paragraph for each form of renewable energy, creating an answer that is clear and well planned. Structuring your answer in this way will allow you to see what information you have already included, helping to **avoid repetition.**

Examples
Throughout the answer, the candidate has **regularly linked back to the question** through the evaluation of different renewable approaches and reference to a number of different countries.

Detail
The candidate has not only demonstrated their knowledge of the different countries that have adopted renewable energy approaches but they have also provided detail of the conditions required for each approach, with an explanation of how power is generated.

Glossary

Geothermal energy: Thermal energy generated and stored in the earth.

Hydroelectric power: Generated from the potential energy of dammed water driving a water turbine and generator.

Renewable energy: Energy that comes from resources that are naturally replenished and can be used again and again.

Solar energy: The conversion of sunlight into electricity.

Tidal power: Electricity generated from the movement of tidal flows.

Wave power: Electricity generated from the movement of waves.

Wind power: Energy that is derived from wind through the use of wind turbines.

SECTION 14

Geographical Skills

Application of Geographical Skills

Section 4 is made up of one question worth ten marks. The question will consist of a scenario that will be accompanied by a number of sources including an OS map, graphs and diagrams. In order to answer the question fully, you should refer to all sources provided.

Q The city of Lincoln is holding a 10k race. A route has been proposed working to the brief. Study Figures 14.1 and 14.2.
Referring to map evidence and other information from the sources, **evaluate** the suitability of the proposed route in relation to the brief.
You should suggest possible improvements.

Figure 14.1: Brief for Lincoln 10K race; advertising poster; visitor numbers and route map

The route should:

- be suitable for all participants
- cause minimum disruption to people and business in the local area
- promote business in the local area
- have a suitable start/finish line
- be scenic/interesting for participants.

**Lincoln 10k Race
Sunday 16th February**

Join over 5000 people taking part

Live music from local bands

Safe route along closed roads

For further information on the race and nearby accommodation go to:
www.visitlincoln.co.uk

Visitor numbers to Lincoln

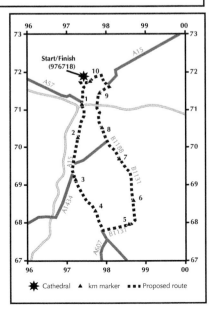

133

Figure 14.2: Map of Lincoln

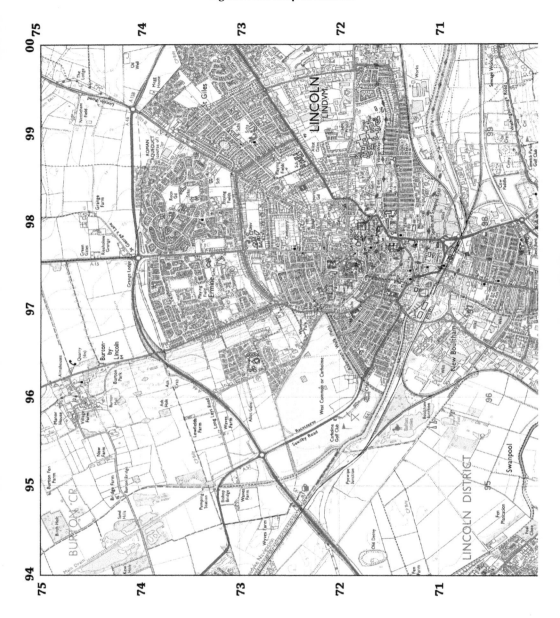

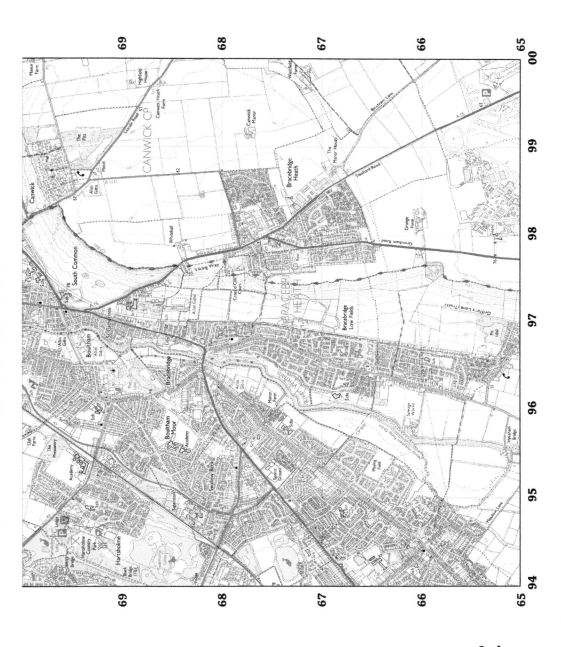

Scale
1:40 000

Break down the brief

Before you start to write your answer, take time to study the brief and sources provided. Work out exactly what the brief is asking you to do and consider how the sources will help you to arrive at a conclusion. It is useful to use the brief when structuring your answer. This will help you to develop your answer and ensure you hit all necessary points.

Below is an example of how this brief may be broken down, giving you an indication of the points you may want to consider.

Suitable for all participants

- Look at the relief of the land: is the route flat/steep? Does it provide a variety for runners?
- Look at the terrain: will the whole route be completed on tarmac or will runners be expected to run off-road?
- Think about weather conditions: if the weather is bad on the day will runners be fully exposed or will the route offer some protection?
- Is there adequate space available throughout the course to accommodate all runners/spectators?

Cause minimum disruption to people/business in local area

- Does the route travel through any main shopping streets/residential areas?
- Are any major roads affected?
- How will an influx of visitors affect litter/noise level?

Promote business in local area

- How will businesses benefit from an influx of visitors?
- Is February a good time to hold this event?
- Have a wide range of businesses been taken into consideration when planning this route?

Have a suitable start/finish line

- Think about accessibility?
- Is there suitable car parking/proximity to public transport?
- Is there adequate space for spectators?

Be scenic/interesting for spectators

- Does the route incorporate a range of environments?
- Are there any points of interest along the route?
  - i.e. rivers/canals/historical sites or monuments, etc

Suggestions for improvement

- How could accessibility be improved?
- How could congestion be reduced?
- Is there a more suitable route nearby?
- Could residential areas be avoided?
- Would the event be more suited to a different day/time of year?

You may want to scribble some key points around the sources provided. This will help keep you focussed as you write your answer.

Answer A

Suitable for all participants

At the start line, the streets and roads are narrow and so can result in a 'bottle-neck' at the beginning of the race. This could cause injury or result in runners finishing the race in a slower time than anticipated. **(1 mark)** *There is variety along the way, with a steep uphill section on the A15 followed by an even steeper downhill section on the B1188. This adds interest to the route. However, this could cause difficulty for some runners.* **(1 mark)** *The route is also very exposed between the 4km and 7km marks, which can cause some difficulties for runners if there are poor weather conditions on the day.* **(1 mark)** *However, the route is generally flat, encouraging runners of all ability.* **(1 mark)**

> ## Top Tip ✓
>
> All information has been given for a reason. Therefore put yourself in the position of those in the brief – what is it you would be looking for/concerned about, either as a participant, a businessman, a spectator or a local?

Cause minimum disruption to people and business in the local area

The route goes through a residential area in grid squares 9769 and 9770. Road closures on race day could limit access for the local people. **(1 mark)** *The route also uses the A57 (dual carriageway), which could also increase congestion on smaller roads and cause disruption to those living and working within the area.* **(1 mark)** *With over 5000 people taking part, locals may see an increase in litter levels as well as experience an increase in noise level from the live music along the route.* **(1 mark)** *However, having the run on a Sunday should minimise disruption as the area/businesses are likely to be quieter.* **(1 mark)**

Promote business in local area

With reference to graph Q12, February is the month with the lowest number of visitors to Lincoln, with just 1 million. Holding the race at this time will bring in extra money to the area at a time when trade is traditionally low. **(1 mark)** *The route starts and finishes in the CBD of Lincoln, promoting the local area to runners/supporters.* **(1 mark)** *With over 5000 runners and their supporters, local shops will benefit, particularly those close to the start/finish line.* **(1 mark)** *Hotels, B&Bs and camping/caravan sites will also benefit, with a large number of visitors requiring accommodation.* **(1 mark)**

Have a suitable start/finish line

There is a lack of space in the CBD and so car parking close to the start line may be problematic. **(1 mark)** *However, the start/finish line is close to a train and bus station (976709) allowing some participants to arrive by public transport.* **(1 mark)**

Be scenic/interesting for participants

The route starts and finishes in the historic centre of Lincoln where there is a castle (972719) and cathedral amongst other attractions for runners to look at. **(1 mark)** *The urban and rural environments along the route will also provide variety for participants, maintaining interest along the way.* **(1 mark)**

Possible suggestions may include

Taking the race out of the city centre could allow easier access by car, for example the race could start in Hartsholme Country Park/West Common Race Course. **(1 mark)** *Hartsholme Country Park is very scenic with large areas of green space, forestry and lakes. Changing the date of the race to the summer months may also encourage more runners to take part, as visitor numbers at this time are at their highest (over 1.8 million in August). This will further increase trade in the city.* **(1 mark)**

10 out of 10 marks

What makes this a good answer?

This is a very detailed response where the candidate has used the brief to help structure their answer. Full analysis of all sources has resulted in an answer that is in-depth and successfully evaluates the route in relation to the brief. By directly lifting information from the sources in the way of figures, descriptions, place names and grid references, the candidate is able to produce conclusive statements, providing suggestions on how the route can be improved.

Now compare this answer with answer B. What differences do you notice?

Answer B

This route is good for local businesses as the influx of people will bring in more money to the area and people will need somewhere to stay. **(1 mark)** *The route is flat with some hills. There will be* *some parking spaces available because the race takes place in the city and it is usually easy to get parked there. There might be trains nearby too, which can be useful. In 6996 the runners will run through a housing estate. This might annoy people. If a lot of people are running at the same time the city centre could become very crowded and people might get angry.*

The steep land may be difficult for some people and so this may discourage people from taking part. However, this is a good route for runners as there is lots to see and do in Lincoln. There is a golf course, many museums and lots of churches. It will also suit hotel owners, as runners will need somewhere to stay. They may also choose to stay in surrounding areas like Nettleham (which will be cheaper than staying in the CBD), therefore boosting the local economy. **(1 mark)** *The high number of visitors may cause an increase in rubbish and noise pollution. As the route passes through residential areas, this will annoy local people, particularly on a Sunday when people may want to rest and have a quiet day.* **(1 mark)**

3 out of 10 marks

What makes this a weak answer?

The first sentence is a developed point so gains **1 mark**. However, the rest of the answer is largely descriptive with little reference to the sources provided. Although the candidate has included a grid reference, this is inaccurate and the other information is largely irrelevant. In addition, when the candidate makes reference to 'steep land', they include no map evidence to back this up.

Poor structure has resulted in an answer that is vague and repetitive. The candidate has only referred to a few of the points in the brief therefore has failed to fully analyse the suitability of the route. Poor analysis has prevented this candidate from making suggestions for improvement, limiting the number of marks that can be awarded. However, the candidate has managed to pick up **two marks** from the last three sentences, which have been developed, resulting in **three out of ten** overall.

General advice on using sources

- When giving map evidence, include descriptions, grid references and/or place names. A vague, over-generalised answer that makes no reference to the map will be awarded no more than **4 marks**.
- Always provide 6-figure grid references unless you are referring to something that covers the whole grid-square, e.g. a large settlement or area of forestry.
- Think about structure – use the brief provided to help you structure your answer by taking each factor in turn. For each factor, you should refer to at least 1 source.
- Throughout your answer, you should make reference to the sources and offer an explanation with reference to the brief. An answer consisting solely of limited descriptive points will gain a maximum of **5 marks** out of an available 10.

Maximising Marks

In order to gain 1 mark, you should make reference to the resource before giving an explanation with reference to the brief, e.g.

'The route uses the A57 (reference to the resource). *Closure of this dual carriageway will cause increased congestion on smaller roads* (explanation) *causing disruption to people and businesses in the local area* (reference to brief).'

OR 1 mark can also be achieved through limited description/explanation of **two** factors, e.g.

'Runners may **spend money** *in local shops as they wait to start the race. They may also* **need somewhere to stay** *if they have travelled far, and so local hotels will benefit.'*

Q Study the following resources

- Map A: Location of Chelson Meadow landfill site (figure 14.3)
- OS map B: (Extract 2006/OL20: Plymouth) (figure14.6)
- Wind Rose C: Plymouth (figure 14.4)
- Information D: BBC Headline (figure 14.5)

Figure 14.3: Location of Chelson Meadow landfill site

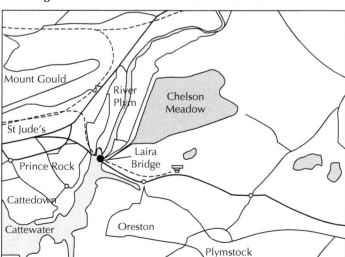

Figure 14.4: Wind Rose C: Plymouth

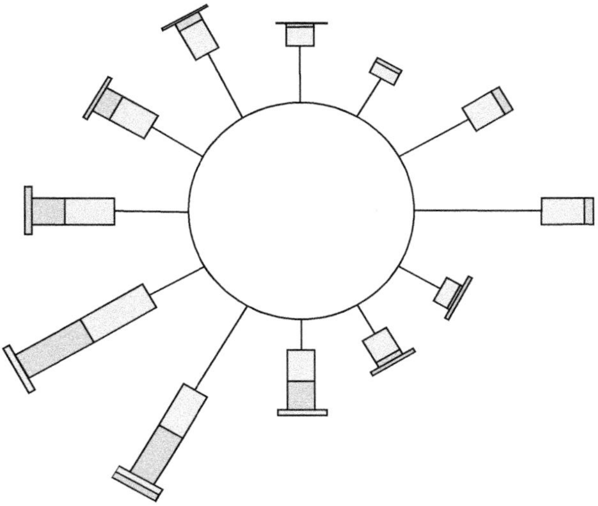

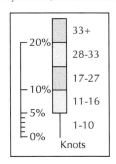

Wind rose for the period 1991-2000
at Plymouth (50 metres a.m.s.l.)

	Knots
20%	33+
	28-33
10%	17-27
5%	11-16
0%	1-10

Figure 14.5: BBC Headline: August 2011

"Plymouth's Chelson Meadow tip restored to grassland"

Plymouth City Councillor Michael Leaves said the restoration of the site would "make a huge difference to the people who live nearby", the work to revert the site to its natural state cost £18m. Domestic waste is now transported 20 miles away to a private landfill in Liskeard, Cornwall.

Figure 14.6: OS Map B (Extract 2006/OL20: Plymouth)

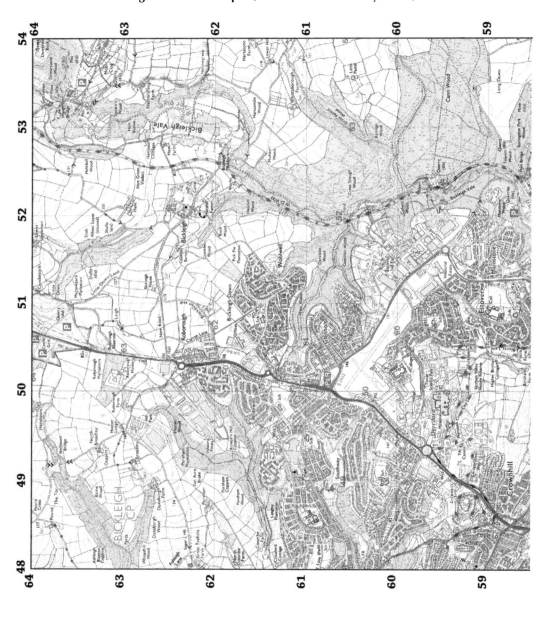

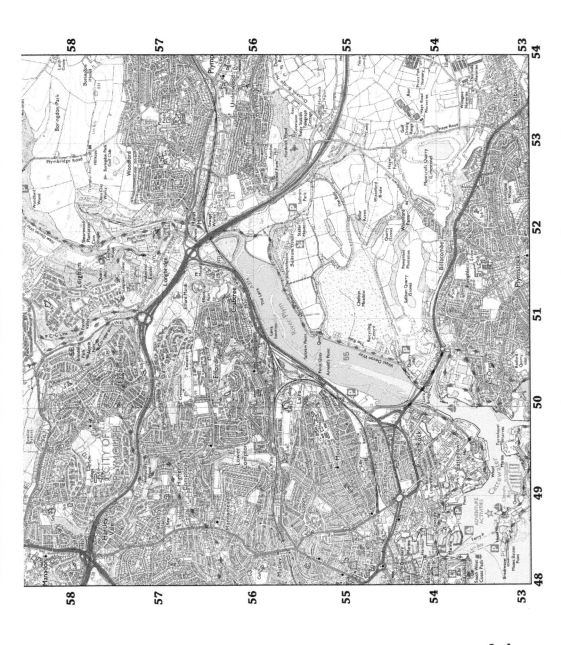

Scale
1:40 000

(a) Discuss the suitability of Chelson Meadow as a site for landfill.   **5 marks**
(b) Evaluate the social, economic and environmental impacts of the closure of this site.
                                                                          5 marks

Breaking down the question – part (a)

Before you start to write your answer, you must first think what the question is asking you to do. If you are asked to **discuss** *the suitability* of something you are expected to consider a range of different views on a given issue. Your answer does not need to be balanced but it is a good idea to refer to both positive and negative factors in order to add depth and value to your answer.

Therefore, if asked to **discuss** *the suitability of a site for landfill,* you must first consider what developers are looking for and who/what might be affected by the proposal.

What should you consider when selecting a site for landfill?

- Look at the relief of the land. Is it flat? Easy to build on?
- Price of land: is the site close to the city centre or in the cheaper suburbs?
- Is the site easily accessible? Are there good road networks nearby?
- Is the site big enough to accommodate an increase in demand for landfill? Is there room for expansion if required?

Who/what might be affected by the proposal?

- House owners: property prices could decrease.
- Local ecosystems: water sources could be affected.
- Is the site close to centres of population? Will local people be subject to visual/air pollution from the site?

> ## Top Tip ✔
>
> Once you have read the question and analysed the sources, read the question again. Scribble some points around all sources provided to help keep you focused and keep your answer relevant.

Answer A

I think Chelson Meadow is a suitable site for landfill. It is a good site because it is not close to housing. People who live in Plymouth may object to a landfill being close to their house. However, Chelson Meadow is not near housing so it won't bother anyone. The land is quite steep so after a while the rubbish heaps may become too heavy and slide into the river. This could harm fish stocks. There is a main road next to the site and this will make it easy for lorries to access the site, bringing waste from the city. **(1 mark)** There is a recycling centre close by and lots of forests. The people living nearby in Plymstock may be unhappy with the rubbish dump being next to their house.

1 out of 5 marks

What makes this a weak answer?

Answer A is poorly structured and contains little detail. The first four sentences are repetitive and add little value to the answer. The candidate makes a valid point with reference to the main road and accessibility. As this point has been developed, 1 mark is awarded. However, this could have been enhanced through the use of map evidence (e.g. the name of the road). Throughout the candidate makes very little reference to any of the sources provided, resulting in an answer that is vague and over-generalised. Brief statements like *'There is a recycling centre nearby and lots of forestry'* are of little relevance unless further developed. Therefore, vague comments and poor technique have resulted in an answer worthy of just **1 out of 5 marks.**

Answer B

Chelson Meadow is a relatively flat area of land on the edge of the city. As it is far from the city centre, land is likely to be cheaper. **(1 mark)** *The A379 runs approximately 0.5km south of the site, providing good access by road. This is useful for transporting waste from within Plymouth.* **(1 mark)** *As Plymouth is a large city, a large area of land will be required to dispose of waste. Around 1km² is available here, which will allow the landfill to expand as demand is increased.* **(1 mark)** *Visual pollution will be minimal as this site will be screened by forested areas, e.g. Pomphlett Plantation, and the River Ply embankment.* **(1 mark)** *It is unlikely for the land to be used for any other commercial or residential use given the likelihood of flooding due to the close proximity of the mudflats.* **(1 mark)** *As shown on the wind rose the prevailing wind is south-west. This prevents much of the city from being affected by the smell of the landfill. However, suburbs to the east of the city, e.g. Underwood, may be affected. (1 mark)*

5 out of 5 marks

What makes this a strong answer?

Having achieved 5 out of 5, answer B is well structured and regularly links back to the question. The candidate has made good use of the sources available, ensuring the answer is relevant with appropriate levels of detail. The candidate refers to a range of different factors, which adds depth and demonstrates a high degree of understanding. From the level of detail included, the candidate has successfully analysed all sources provided and used the information to support their answer. Overall this is an answer of very high quality and one typical of an A student.

Breaking down the question – part (b)

If asked to **evaluate** something you are being asked to *make a judgement.* Therefore, you should be able to comment on how the closure of the landfill site will affect people, the economy and the environment. Again, these impacts can be both positive and negative.

For full marks to be awarded, you must refer to all aspects of the question. If social, economic **and** environmental impacts are not referred to, your answer will be marked out of 4. Figure 14.7 illustrates some of the factors to which you should refer.

Figure 14.7: Social, economic and environmental impacts

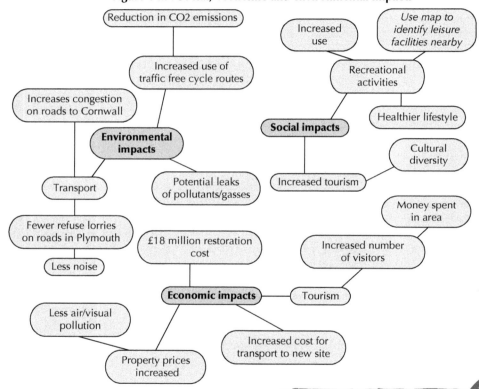

Look at the two answers below. What differences do you notice between the two answers? Think about structure, attention to detail and use of resources.

Answer A

If the site closes, tourism in the area is likely to increase, particularly in places like Saltram House (517556) and Saltram Wood (512555). **(1 mark)**

More people may be inclined to visit the area to cycle or to walk the West Devon Way. This will increase the number of visitors to the area and in turn will boost the economy. **(1 mark)**

Local residents may now be more attracted to use the traffic-free cycle routes. This will not only encourage people to live healthier lifestyles but will also reduce the number of cars on the road, reducing pollution. **(1 mark)**

Residential areas such as Underwood will become a more pleasant place to live as it will no longer be subjected to the smell of waste, nor will residents have to put up with the same degree of visual pollution. **(1 mark)**

The number of refuse trucks using the A375 will decline, reducing congestion and noise for residents, e.g. in Billacombe. **(1 mark)**

The closure of the site could also have a negative impact. The £18 million restoration cost combined with the increased cost to transport waste 20 miles away to Cornwall could lead to an increase in local taxes and a decline in local services. **(1 mark)** Refuse lorries will now have to travel further to the new site, leading to an increase in congestion and transport costs as well as an increase in CO_2 emissions. **(1 mark)**

5 out of 5 marks

Answer B

If the landfill site closes, towns and villages nearby will no longer be smelly. More people may visit the area and fewer people will be inclined to move out of the city as it will again be considered a nice place to stay. **(1 mark)** Landfill sites are not popular with residents. However, as shown on the map, the landfill site is situated at the edge of the city. Therefore, if it closes it will not have much effect as it is away from most people anyway.

Closing the site does not solve the problem as people still produce too much waste. If the site closes it just transfers the problem to another area and another group of residents will be unhappy.

1 out of 5 marks

What answer comes out on top?

Answer A is a very structured response that makes full use of the sources provided. By referring to a range of social, economic and environmental factors the candidate covers all aspects of the question while referring to all sources to ensure each point is fully developed.

Each point follows a similar structure whereby the candidate states the impact before evaluating that impact through analysis of the resources. For example,

'The number of refuse trucks using the A375 (analysis of resource) will decline (impact), reducing congestion and noise for residents (evaluation), e.g. in Billacombe (analysis of resource).'

This technique will encourage the candidate to stay focused and ensure maximum use of time. Overall, this is an excellent example of how to use sources in order to help you arrive at a conclusion and the candidate deservedly achieves full marks.

In comparison, answer B is particularly poor in that points are not developed and information is not relevant. The first paragraph achieves 1 mark for a limited description of two points. Marks could have been gained here through the addition of map evidence or reference to other sources.

The second paragraph is largely irrelevant and adds little value to the answer. Vague statements like '... *will not have much effect as it is away from most people anyway*' are careless and a poor use of time. Similarly the last paragraph is very basic and fails to answer the question. Little reference has been made to social, economic or environmental impacts and the sources provided have barely been recognized. This answer reflects poor technique and achieves just **1 out of 5 marks**.

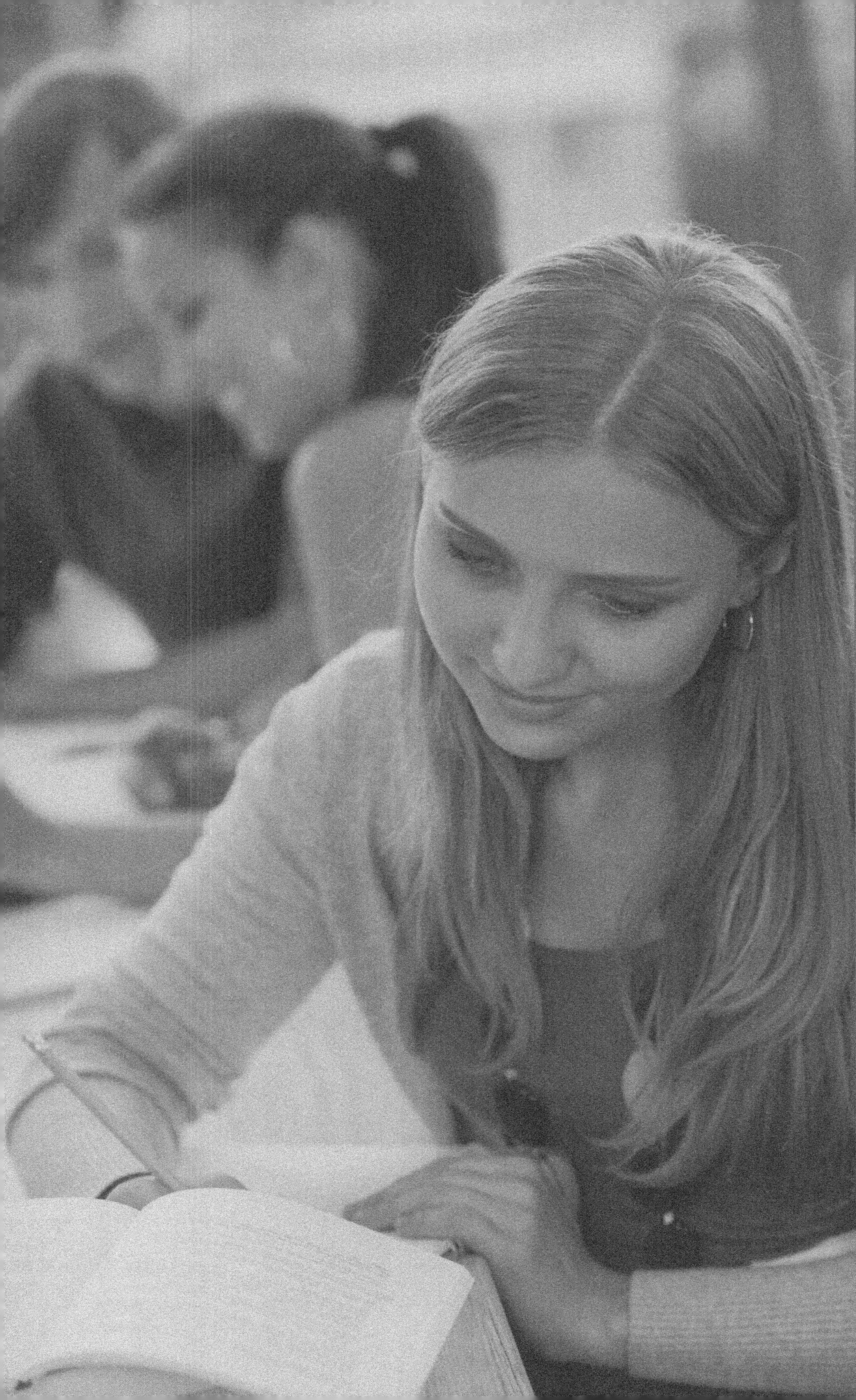

Section 15: Assignment

Assignment

The coursework Assignment section of Higher Geography is marked out of 30 and will be marked externally by the SQA.

The purpose of the Assignment is to demonstrate challenge and application by requiring you to draw on and apply skills, knowledge and understanding within the context of a geographical skill or issue.

The two stages of the Assignment

Research stage

During this stage you should:

- Identify a geographical issue or topic
- Carry out research and evaluate research methods/reliability of sources
- Process/use information
- Demonstrate knowledge and understanding of the topic/issue
- Analyse information
- Reach a well-supported conclusion

Production of evidence stage

This stage involves writing up your Assignment in the form of a report under exam conditions within one hour and thirty minutes. To help you, you are allowed to prepare two A4 single-sided sheets of processed information. This processed information should show evidence of the primary or secondary data that you have collected.

Examples of processed information include:

- Annotated photographs
- Annotated maps
- Field sketches
- Annotated diagrams

- Cross sections/transects
- Production of graphs, charts, statistical tables
- A coded interview transcript

No marks are awarded for the processed information taken into the assessment. However, in order for you to achieve full marks you must make reference to the results.

The allocation of marks

The table below shows how marks will be allocated.

Carry out research on a geographical topic or issue	Max 6 marks
Use of and reference to processed information	Max 4 marks
Drawing on knowledge and understanding of the geographical topic or issue	Max 8 marks
Analysing information	Max 6 marks
Reaching an overall conclusion, supported by range of evidence	Max 2 marks
Communicating information	Max 4 marks

Possible topics

It is important to select a topic that you are interested in and one where information is easily accessible. A selection of possible topics are shown below.

Physical

- Controls on sediment size and shape along a particular coastline.
- Examination of the rate of longshore drift along a particular coastline.
- How do soils vary either side of a moraine ridge?
- An analysis and classification of a particular soil profile.

Human

- Distribution of migrant workers within an urban area: implications for housing, transport and community integration.
- Opportunities and challenges for generating energy in rural Scotland.
- Preservation vs conservation: managing environments along dynamic coastlines.
- Nature and impact of a particular urban development on the local population.
- Future of greenbelts: protecting rural land or hindering urban development?

Global

- Risk to local communities of predicted climate change.
- Mapping inequality within Scotland.
- Impact of globalisation on local/independent shopkeepers.
- A study of the effectiveness of Scottish/UK international initiatives.
- Benefits/problems of a local energy development (e.g. wind farm/power station).

Combine human and physical

- How land-use impacts on the response of a river to a rainfall event.
- How soil characteristics vary as a result of human land-use in rural areas.
- How the expansion of an urban area is affected by physical landforms, drainage and soil type.

Top tips

Get off to an early start

Starting your Assignment early will enable you to gather plenty of information. This will also allow you to be selective. Leaving everything to the last minute will limit the quality of data you are able to collect, affecting the quality of your report.

Choose your topic wisely

It is important that you choose a topic that allows you to carry out fieldwork and research in a variety of different ways. It may be a good idea to think of a 'big question' that you can break down into a series of smaller questions.

Keep it simple

When selecting your topic it is important to keep it manageable. If the topic is too big or the question too complicated, you will end up tying yourself in knots.

Plan ahead

Take time to plan and research your topic before you start. Make sure that a wide range of resources are available and that you are able to gain access to everything you need.

Keep a diary

It is useful to keep a note of everything you do. For example, note down who you have spoken to, what websites you have looked at and where you have visited. It is also a good idea to note down any problems you encountered along the way.

As part of your report, you will be expected to evaluate the usefulness or reliability of the research methods used. Keeping a diary will help you keep track of your work, making the evaluation process simpler.

Work together

Although this is an individual research Assignment, it may be useful to work with others when you are collecting your information, for example, when completing a river study. This will allow you to gather a wider range of information, giving your project more depth.